T0172873

Engineering Graphics Principles
With
Geometric Dimensioning and Tolerancing

E. Max Raisor, FIAE, Professor
Mechanical Engineering
Brigham Young University

Special Acknowledgement

Special recognition needs to be given to Andrew Anderson for his contribution in bringing this work to a wider audience. He has devoted considerable effort to organizing the CD and workbook into its final format.

This project received additional support through the helpful comments and input from Marie Planchard, Budd Langley and Tom Vanderloop. Their assistance is acknowledged and appreciated.

Recognition must also be given to Mary Schmidt, for her gifted editorial work, technical expertise, and cheerfulness on even the most trying days. In addition, this project would have ground to a halt without the nonstop support from Angela Cruse, Andy Cruse, Chris Baker, Mary Nelson, Jeanne Cruse and Keri Schulteis.

Finally, we all owe a debt of gratitude to the man who created it all: E. Max Raisor. He has devoted years to developing and grooming this material. His work marks a new era in engineering graphics.

Stephen Schroff

Trademarks and Disclaimer

The publisher and author makes no representation of warranties of any kind nor any such representations implied with respect to the materials set forth herein, and the publisher and author takes no responsibility with respect to such materials. The publisher and author shall not be liable for any special, consequential, or exemplary damages resulting, in whole or part, from the reader's use of the materials in this book or the accompanying CD-ROM. Every efforts has been made to provide accurate text.

Microsoft and PowerPoint are registered trademarks of the Microsoft Corporation. Other software applications and parts described in this book are trademarks or registered trademarks of their respective owners.

Copyright © 2002 Brigham Young University

All rights reserved. No part of this publication or the accompanying CD-ROM may be reproduced or used in any form, including graphic, electronic, photocopying, recording, or storage media without the express written consent of the publisher, SDC Publications.

Use of the CD-ROM in any form is restricted to individuals that have purchased the Engineering Graphics Principles with Geometric Dimensioning and Tolerancing book. Its use in a classroom setting is restricted to classes for which the book is required of all students enrolled in the course.

Schroff Development Corporation
www.schroff.com

About the Author

E. Max Raisor is a professor of Mechanical Engineering at Brigham Young University in Provo, Utah. Prior to accepting an appointment at BYU in 1968, he had been employed in aerospace and private industry for over ten years. He has taught engineering communications, design graphics, and written and taught CAD courses since the early 1970's. He also pioneered the curriculum development of CAD/CAE systems integration, automated systems management, and productivity measurement courses.

As a practitioner in aerospace and private industry for over 10 years, and as an engineering educator and consultant, he has personally witnessed the evolution of this system of geometric definition and control. He is both emphatic and vocal in his view that ". . .geometric dimensioning and tolerancing is one of the most significant developments in the history of mechanical design. Equal in significance is the fact that when used properly, tremendous benefits accrue in disciplines of process planning, product manufacture, and specification verification or inspection—the results of which are predictable assembly and performance."

Table of Contents

PREFACE

Engineering Graphics Principles with Geometric Dimensioning and Tolerancing embodies a practical experience, and introduces a new approach for learning the basic concepts and applications of engineering graphics, and geometric dimensioning and tolerancing. The materials in the book and the accompanying CD-ROM comply with newest release of the national standard, ASME Y14.5M-1994, and use various instructional methods and tools such as photography, animations, video clips, etc., to heighten their effectiveness as a teaching and training tool.

There are no prerequisites for using these materials. If you have had manual or CAD graphics, or machine shop background in high school, college, or in an industrial application, you may find some of the introductory materials repetitive and less challenging. However, because all of the materials parallel the new ASME Y14.5M - 1994 standard, there is a strong possibility that some portion of the information will be new and challenging to you.

Engineering Graphics Principles with Geometric Dimensioning and Tolerancing has content in the form of a book and CD-ROM. A major portion of the instructional material is included on the CD-ROM. The materials include exercises and self-evaluation materials. Examinations on each of the topic areas are provided in an available instructor's manual.

CD-ROM

A significant portion of the instructional materials provided with *Engineering Graphics Principles with Geometric Dimensioning and Tolerancing* are included on the accompanying CD-ROM. The CD-ROM includes eleven units of application studies, complete with specific examples, problems, support topics, and individual unit examinations. The CD-ROM consists of 3,000 slides, and was developed to assist engineers, designers, and drafters in their pursuit of increased understanding of engineering principles, and geometric dimensioning and tolerancing application.

More than anything else, these materials deal with modeling and defining part geometry on mechanical components and assemblies. Everything in the materials point to the importance of appropriate specification and control of part geometry, including: dimensioning, tolerancing, and verification of geometric characteristics and relationships.

It is the author's view that successful mechanical and manufacturing engineers, designers, drafters, and other technical support personnel, must understand the basic processes involved in the development of graphic representations - whether the documentation is produced manually, or by means of a CAE/CAD system. Like most other modern engineering graphics courses, it begins with an introductory presentation with an abbreviated overview of the historical significance of graphics as a means of conveying and documenting design iterations and modeled configurations. However, it stops short of philosophizing on the design process and cycles of design iteration, and consider only a brief synopsis of the evolution of graphics—primarily from the standpoints of modeling, dimensioning and tolerancing and verification of component geometry.

Four specific modules have been designed to assist the reader in the introduction and/or review of basic concepts related to elements of view selection, orientation, projection, and standards for defining geometry. These units focus on orthographic projection, auxiliary views, sectioning, and coordinate dimensioning and tolerancing.

Practice problems

For each unit of study, there is a series of *practice problems* that are presented on the CD-ROM and coordinated with pages in the book. They are prepared and organized by problem number. As you proceed through the presentations you will encounter slides that refer you to a specific problem (by number) work sheet that is included in this book. Most of the problems have been provided a work sheet to eliminate the need to draw or sketch the initial views of parts. You may be requested to complete the projections, add any and all missing lines, include specific tolerance or dimensional controls, and in other ways verify that the drawing is complete. When you have finished each exercise, you are invited to advance to the next slide in the presentation series, which gives a set of guided instructions, leading to a proper solution.

You are strongly encouraged to complete each of the practice problem sets to assure yourself that you do in fact, understand the basic principles involved.

Self-Evaluation Problems

In addition to the practice problems—with their appropriate solutions, the first four units contain specific *self-evaluation problems* that are at the end of each unit. The design of these assignments is similar to the sample problem practice exercises, but there is no solution slides provided within the topic presentations. The self-evaluation problems are labeled, and instructions for completing them are included, both in the presentations, and on the work sheets themselves. They have been included in the packet to assess your understanding of the principles under discussion. In an instructional program, students may be required to submit the self-evaluation to the instructor.

Instructor Guide

The Instructor Guide for *Engineering Graphics Principles with Geometric Dimensioning and Tolerancing* provides additional support materials that can be used in conjunction with the book and CD-ROM.

Examinations

The Instructor Guide includes eleven examinations (one for each of the instruction modules). The tests are comprehensive for the specific unit being studied, but they may also include the application of principles studied in previous units. Readers are encouraged to spend some quality time with the exams. Some of the questions will likely seem vague and tricky. They have been included to help the reader implement the process of design thought—not to trick the reader into choosing incorrect responses. The examinations are arranged in the following order:

Examination #1	Orthographic Projection
Examination #2	Auxiliary Views
Examination #3	Sectioning Conventions
Examination #4	Coordinate Dimensioning & Tolerancing
Examination #5	Introduction to GD&T
Examination #6	Form Tolerances
Examination #7	Datums and Datum Targets
Examination #8	Orientation Tolerances
Examination #9	Location Tolerances
Examination #10	Profile Tolerances
Examination #11	Runout Tolerances

The examination sheets are provided in the instructor's manual.

AUTHOR'S NOTE

Success doesn't come from doing what one likes, but rather, from liking what one has to do. Adjust your thinking to accommodate all elements presented. Avoid identifying with only selected concepts or elements. Liking things is a learned quality, and happy is the person who can incorporate all of the necessary elements that contribute to true education and career preparation. Planning is the silent ingredient in motivation that stimulates action and makes things happen. You cannot always control life's circumstances, but you can control your attitude towards the outcome of many of life's experiences—if you plan and work effectively. So, have a plan of action, know beforehand what you want to accomplish, and schedule your time accordingly. Plan your work, and work your plan. It is much easier to keep up than it is to catch up. Discipline yourself in the appropriate procedures that will keep you current in all of your work and studies. Keep in mind that if you fail to prepare, you prepare to fail.

The material covered is not extremely difficult, but it is complex. Because of its complexity, some have thought it to be harsh and unforgiving. It need not be so, but may best be thought of as a language. We will first examine its symbolism, rules, structure or semantics. Then we will learn how to write, using its special grammar and graphic symbolism—the international graphics language—to establish and communicate specific information on engineering drawings; information that is uniquely orchestrated to permit only a single interpretation.

It is not the purpose of these materials to respond to all of the diverse possibilities for constructing descriptive views with supporting dimensional and tolerancing data. Likewise, it is not the intention of the author to imply that there are simple solutions to all graphics related problems, or to suggest that the subjects listed earlier are to be mastered without significant effort. However, it is the objective of both the author and these materials, to 1) provide perspective—relative to the significant position that graphics occupies in the design process—and then 2) to provide a measure of carefully articulated and illustrated examples to assist you in understanding the application of symbolism and special tools that are now available to engineers and designers, for the purpose of defining and controlling geometry. The basic rules and applicable concepts will be presented, using simple geometric examples. The generalization and transferal of these basic principles to more abstract and complex applications will become a logical process for you as you gain experience.

EMR

INTRODUCTION

As stated earlier, the major objective to be served in the introductory modules (1-4) of this material is to gain some measure of proficiency in basic graphic skills applications –all of which are focused on correctly and completely defining part geometry. Questions pertaining to the number and layout of required views, special view requirements and specific view orientation, spacing of required views, and dimensioning and tolerancing issues, must be considered. The enhanced possibilities for greater control of feature characteristics and feature relationships follow in the subsequent seven modules.

While much could have been written regarding drafting, the content of the book and CD focus on basic introductory concepts found to be of greatest value to designers and engineers in their graphics communication requirements in the field. The content is based on the philosophical position that engineering graphics is much more than graphic representations of mechanical components and assemblies created by manual drawing or CAE/CAD modeling techniques. Engineering graphics incorporates manual or computer-assisted model generation and description, but it must also communicate all of the specific dimensioning and tolerancing data required to successfully manufacture, inspect, and implement the component in its assembly. Properly understood, and complete in descriptive specification, engineering graphics is, in reality, the international language of design, manufacturing, verification and assembly. It is the language by which critical data is communicated across international boundaries, worldwide. The main goal is to provide for each reader, the grammar and interpretive tools of this international symbolic language. The following are the basic objectives of the book's contents:

- Develop a working understanding of engineering graphics as a means of interpreting and communicating design intent through the use of internationally standardized methods and symbolism (all geometric modeling will be done using third-generation CAD systems).

- Develop reading and technical sketching skills to assist in correctly interpreting and graphically solving problems related to engineering graphics, including geometric constructions, projection theory and applications, plane and descriptive geometry, sectioning conventions, and coordinate dimensioning and tolerancing.

- Become familiar with current national and international graphics standards governing dimensioning and tolerancing applications (ASME/ANSI Y14.5M-1994 and ISO). Apply dimensioning and tolerancing specifications on geometric models, using standardized symbolism, for uniform application and interpretation of coordinate and geometric dimensioning and tolerancing (GD&T). (Limited consideration will also be given to manufacturing and verification methodologies.)

Ancient wisdom suggests "the palest ink is better than the best memory". Creative ideas are occasionally lost because insufficient effort is made to preserve them. Within the engineering enterprise, conceptual and iterative sketches, and drawings are the principal sources of tracking technical milestones in the process of engineering solutions to technical problems. In order to correctly preserve and present this process, accurate and complete detail drawings and/or CAD models must be created. These graphic representations must be clear in their meaning, and standard in their descriptive technique. They must allow for the maximum amount of part/feature variation without compromising part function, and they must provide assurances that the part(s) will function as intended.

Graphics standards are the key to producing drawings of predictable quality. Information on engineering drawings would have little value if there were no standards against which the information could be measured and evaluated. Standard procedures in creating representative drawings must include the modeled geometry—complete with appropriate descriptive views, and comprehensive dimensioning and tolerancing data. The dimensioning and tolerancing information must be complete, with maximized tolerance values. The current (U.S.) national standard for dimensioning and tolerancing is ASME Y14.5M-1994. It will become the focus of our study during the last half of the course. The document is the result of continuous action to establish, where possible, agreement with the global standard produced by the International Organization for Standardization (ISO).

Some updates and changes that have occurred since the 1973 and 1982 ANSI Y14.5 standards were released will be highlighted and explained. This may prove to be the more challenging portion of the materials.

Historical Background

Every textbook I have ever read on the subject of engineering graphics, begins with a statement (or chapter) of historical background; its development over the years, special periods of change, the dawning and impact of CAD, etc. I have included a brief historical sketch of my own—touching only on the high points of the actual evolution—with emphasis on the development of dimensioning and tolerancing. However, in deference to the research efforts of those who have provided much more information than I am inclined to do, I will forego the temptation to repeat what they have written, and simply recommend that if you have an interest that carries you beyond the extents that my brief statement provides, you may obtain a copy of virtually any college-level textbook on engineering graphics or drafting, and read the opening statements with regards to the evolving historical events.

The evolution of engineering graphics (dimensioning and tolerancing in particular), prior to and during the earliest stages of World War II (late 1930's through the early 1940's), was sluggish at best. Some initial interest in establishing an American Standard for engineering graphics was expressed in the early-to-mid 1930's, but that effort, (ASA standard of 1935), was woefully deficient by today's standards, and was essentially ignored until midway through WWII, when major changes were necessitated by the needs of a changing wartime society and a competitive industrial economy. —But wait, I'm getting ahead of my story. Read on only if you choose to do so.

This aggressive, modern movement has been underscored by increased precision requirements in both design and manufacturing. Thus, in recent years, increased emphasis is being placed on cooperation and team service in design and manufacturing. Historically, the need for associating tolerances with dimensions became apparent as the demand for precision increased. During and immediately following World War II, engineering graphics (specifically, dimensioning and tolerancing) rounded a significant corner. This period of time was the turning point for engineering graphics standards among all of the industrial nations of the world. The United States was becoming recognized as a leading manufacturing entity in the world economy, and competition was driving production standards. Engineers, understanding that variation was unavoidable, became more involved in the design process by evaluating the need for tolerances associated with every dimension—on every part, in every assembly. What was equally important was their awareness that variation was acceptable, as long as it was contained within specified limits. Thus, the concept of plus and minus coordinate tolerances emerged, and was integrated into engineering drawings.

The British led out in creating standards that would eventually be the seedbed for what is known today as true position tolerancing. Stanley Parker, an English worker in a torpedo factory in Scotland, devised a method of specifying cylindrical tolerance zones surrounding an absolute location, from what was previously specified as rectangular 'plus and minus' tolerances. Thus was born the concept of true position, leading to interchangeability of parts within assemblies.

A few years earlier, the American Standards Association (ASA, 1935) published the first U.S. standard for engineering drawings: the *American Drawing and Drafting Room Practices*. It was a very brief document, having fewer than 20 pages in the complete publication. There were only 5 pages in the booklet that dealt with the subject of dimensioning, and its coverage of tolerancing was limited to 2 brief paragraphs. Inadequate at best, it was the beginning of what would eventually become our current national standard for dimensioning and tolerancing on engineering drawings.

A publication called the *Draftsman's Handbook* was published in 1940 by the Chevrolet division of the General Motors Corporation. This publication incorporated some basic concepts related to position tolerancing. It was followed in 1945 by the U.S. Army publication of its *Ordnance Manual*, which specified dimensioning and tolerancing standards for the Army. It was in this manual that symbols were first introduced in the U.S., as a substitute for long hand notes in defining and controlling position tolerances.
When the second edition of the *American Drawing and Drafting Room Practices* was published in 1946, its content had been expanded, but the specific information on tolerancing was still woefully inadequate to meet the increasing requirements for accurate definition of geometry and geometric relationships. During the same year (1946), the Society of Automotive Engineers (SAE) revised its SAE Aeronautical Drafting Manual, recommending dimensioning and tolerancing enhancements.

Military Standard 8 (MIL-STD-8) was published in 1949. Its purpose was to specify dimensioning and tolerancing standards for the military. It was subsequently revised and expanded in 1953. In this revision, seven geometric characteristic symbols were identified for the purpose of defining and controlling tolerance zones for physical parts geometry, and geometric relationships. In essence, a new philosophy was established in this revision, to accommodate functional dimensioning methodologies.

Beyond the obvious difficulties of having inadequate and inconsistent specifications for dimensioning and tolerancing practices on engineering drawings, there was no standardization between the independent groups that were publishing the standards. There were major differences in the philosophies endorsed by each of the military, government, and industrial entities, and there was little interest in structuring a mediated agreement between the participants. The ASA, SAE, and Military, each independent of the other, argued to defend their posture on their unique interpretations of what the standard should be.

After years of inconsistency and debate, in 1957, the American Standards Association Standards Institute (ASASI)—successor to ASA, approved the first American standard devoted to dimensioning and tolerancing. The development of the standard was accomplished in cooperation with the Britain and Canada.

In 1959, the last revision of MIL-STD-8 was released (MIL-STD-8C), which brought the military much closer to the ASA and SAE standards. In 1966, the first unified standard was published by the American National Standards Institute (ANSI)—successor to ASASI. The standard was known as *ANSI Y14.5, Dimensioning and Tolerancing on Engineering Drawings*. The standard was updated in 1973, and again in 1982. The newest release of the standard was written in 1994 and published in 1995. Written by the American Society of Mechanical Engineers, the present publication is known as *ASME Y14.5M-1994* (the 'M' denotes metric). Parties involved in the international committees and sub-committees, chartered to unify graphic standards world wide, report that *ASME Y14.5M-1994* is approximately 95% - 97% in agreement with the ISO global standards for dimensioning and tolerancing. Some of the remaining differences will be explained as we progress through the course.

Measuring Unit Values: A Final Word

There are only two measuring unit values that are used extensively today; the decimal inch, and the metric system. Most countries throughout the world use the metric system. The standards do not discriminate between the two, so it makes no difference which system is used. This is because the methods of tolerance calculation do not differentiate on the basis of the unit of measurement employed. Many examples of both systems of measurement will be provided in future presentations dealing with geometric dimensioning and tolerancing.

The Dynamics of Change

Consistent with the changes that have occurred in dimensioning and tolerancing, just about everything else around us—at least in the technical world—is also evolving through new cycles of development. Because we are caught up in a world of change, I thought you might enjoy what is being said by some of the world's foremost authorities on the subject. It's brief, but interesting. You may wish to ponder the implications of some of the statements. If not, you may move onto the beginning of the first topic –Fundamentals of Orthographic.

"No transformation produces a static and lasting result. Hence our environment is at any moment of human history the product of . . . a continuous process of change. Man's adjustments to his environment are not a series of unrelated stages of development, . . .but an integrated chain of events. Thus, permanency exists only in the uninterrupted continuity of change and in the dynamic relations among all aspects of human activities."

E. A. Gutkind

"Within a decade or two it will be generally understood that the main challenge to U.S. society will turn—not around the production of goods, but around the difficulties and opportunities involved in a world of accelerating change and ever-widening choices. Change has always been a part of the human condition. What is different now is the pace of change, and the prospect that it will come faster and faster, affecting every part of life. The movement is so swift, so wide and the prospect of acceleration so great, that an imaginative leap into the future cannot find a point of rest."

Max Ways

"In almost every application of development one may wish to measure, there is a constant acceleration; the changes become larger and they occur more frequently as we move forward in time."

Kenneth Boulding

Several years ago, Scientific American plotted Adams' "Law of Acceleration". Graphs were made of such processes as the discovery of natural forces, and the time lag between each successive discovery. Tables were made that plotted the isolation of natural elements, the accumulation of human experience, the speed that transportation has achieved from the pace of man walking, to space satellites, and the number of electronic circuits that could be put into a cubic foot of space. In every case the rising curves on the graphs showed almost identical shapes, starting their rise slowly, then sharper and sharper until, in our times, nearly every trend of force is embarked on a vertical course.

"The worldwide explosion of technology, which during the past 30-plus years has generated 90% of our current knowledge in engineering and the physical sciences, has greatly accelerated. Bruce Marrifield, U.S. Assistant Secretary of Commerce [1985], predicts more technological change in the next 10-20 years than has happened in all of history combined."

The Anderson Report, VIII:I
September 1985

"There is not much we can do about the past, and the present is so fleeting, as we experience it, that it is transformed into the past as we touch it."

Dostoevsky

"Twenty-five percent of all the people who ever lived are alive today; ninety percent of all the scientists who ever lived are living now, and the amount of technical information available doubles every seven years."

Alvin Toppler [1967]

"Many otherwise capable people are unable to look ahead or to contemplate those future circumstances that come outside their own experiences or the circle of their present needs. When one has his eyes focused too much on the present, the future tends to be blotted out of vision. Vision accounts for some of the greatest differences between professionals. The person who has vision thinks about and prepares for the future, while the person who lacks vision lives each day as it comes."

Sterling W. Sill

Overview

Because each unit has its own introduction regarding the significance of the topic to be covered, the following outline statements have intentionally been abbreviated. You are strongly encouraged to carefully read (and ponder) the introductory statements and objectives at the beginning of each module. (These outlines are also available in Appendix 3 of the CD-ROM course presentation). Not only is it a good idea to initially familiarize yourself with the material that each unit will cover, but it seems logical to me, that subsequent readings of the objectives and introduction materials for each unit will help you better understand the concepts being studied.

Orthographic Projection

Orthographic projection is the means of imaging three-dimensional objects on a two-dimensional plane. Specific terms and concepts, integral to the international graphics language, are defined and illustrated. The 'Glass Box' method of visualizing the projection techniques for first- and third-angle projection is illustrated. Object orientation and spatial relationships are discussed, and various applications used to accommodate specific orthographic view projections. Appropriate examples of each are demonstrated.

The significant differences between first- and third-angle projection techniques are explained, and clarification provided as to when and where each would be used. Various applications used to develop view projections are examined, and examples of each are demonstrated. Additional problems and exercises are coordinated between the CD and this packet, and are resolved in step-by-step procedures.

Half and partial view projections are discussed, and standard hardware (screws, bolts, nuts, pins, keys, shafts, etc.) projection techniques are demonstrated. Additional orthographic projection problems are presented and illustrated, including cylindrical surface vanishing edges, and casting and forging run outs. Examples of solution approaches are demonstrated. More problems and exercises are presented, and resolved, as a final review of orthographic projection concepts and principles prior to the unit examination.

Auxiliary Views

The need to describe model geometry that is inclined or oblique to the projection plane(s) is discussed. Introductory concepts, principles, and terminology related to descriptive geometry are presented. Drawings requiring primary, secondary, and/or tertiary auxiliary views are resolved by using these basic principles. Typical auxiliary views include the following problem types: true size and shape of planar surfaces—(inclined and oblique), true angles between intersecting planes, true lengths of lines, shortest distance between a point and a line, shortest distances from a point to a plane, etc. The major concepts covered in this first lecture are: 1) inclined and oblique projection planes, 2) true length lines, 3) true size and shape of an inclined plane, and 4) true size and shape of an oblique plane.

This unit explores the need to solve for the true angle between lines or planar surfaces that intersect, or would intersect if extended in three-dimensional space. Appropriate use of the basic rule for true length lines is emphasized. The significance of the line of intersection in solving true angle problems is established and demonstrated. Several examples of true angle problems are provided and solution requirements evaluated.

The unit on auxiliary views also explores the need to solve for the shortest distances between non-intersecting lines or between a point and a planar surface in three-dimensional space. Using the basic rule of true length lines is again emphasized and illustrated through solving several related problems.

Cylindrical features in auxiliary views are also examined in this unit. Primary, secondary and tertiary views of cylindrical objects are developed and explained. The concepts of auxiliary views in mechanical drawings and geometric modeling are reviewed, and questions answered. The significance of auxiliary views in geometric definition and part description is reviewed to demonstrate the importance of modeling and defining objects completely to accommodate manufacture and inspection of controlled features.

Sectioning Conventions

An introductory presentation examines the theoretical concepts of cutting or dividing parts to allow complete descriptions of internal geometry. Frequently, a complete part definition cannot be established without reference to internal variations in geometric configuration. These feature variations may not be visible as the object is viewed externally, such as with parts produced using investment casting. Line conventions, cutting planes, section designators, unique symbolism, and basic section types (full, half, offset, partial and aligned sections), section view placement priorities, standard break symbols and crosshatch symbolism are discussed and illustrated.

A comprehensive review of sectioning conventions is given, including aligned features in section views—such as rotation of spokes, ribs of castings, webs and stiffeners, and standard hardware items in section, such as fasteners, shafts, etc. Longitudinal sections depicting hardware items (bolts, nuts, screws, pins, shafts, keys, etc.) are explained. Cross sections of these same parts are also discussed. Use of crosshatch angle, and standard crosshatch methods are also presented.

Coordinate Dimensioning and Tolerancing

A background introduction to dimensioning and tolerancing is given which includes the philosophy for part design, manufacture and verification. Such specifications require a clear statement and/or definition of the part geometry—be that a detailed drawing, a CAD geometric model, or a specific communication that describes and controls the specifications for the part and its related functions. The unique symbolism (ASME Y14.5M-1994) is introduced, which includes implementation rules for proper application.

Coordinate dimensioning is governed by a set of standard rules and guidelines. The basic rules for coordinate dimensioning are discussed, and the basic steps to successful dimensioning are introduced. Design considerations for part and feature function and relationships are discussed and the concepts illustrated. Feature size and location are shown to be essential elements in part and feature definition, and the significance of dimension origin points are emphasized. Dimensioning techniques, choices, or options available to the designer/engineer are illustrated, and dimension placement issues are also explained. Interpretation of drawing specifications is emphasized.

This unit undertakes an explanation of standard symbolism, guidelines and rules used in dimensioning and tolerancing of mechanical components and assemblies. It also extends into the applications, methods and procedures employed in using these standards to describe and control part geometry. Each of the major applications and basic principles discussed are illustrated with example problems and interactive solutions. The problems are designed to establish in the minds of the students, the need for greater control of end product results to assure useable parts—at minimum cost.

Coordinate tolerancing is introduced with emphasis on the fact that mechanical perfection is non-achievable in the realm of finite realism. Geometric variations in part form, size, and feature location are always present in the world of design and manufacturing. Various techniques depicting the methods of specifying acceptable variation limits in mechanical components are explained. Basic hole and basic shaft system philosophies are introduced, and specific problems solved to demonstrate the applications. Applications of dimensioning techniques to control tolerance build-up are also illustrated and discussed.

The concepts of 'standard fits' between mating mechanical parts is introduced. The use of standard fit tables is explained and their functionality illustrated by solving several critical and non-critical fit problems. The crucial role of tolerancing mating parts is reviewed, with emphasis being placed on the problems related to feature location, feature size, form, and orientation when relying on standard coordinate dimensions and tolerance controls. Datums and datum features are re-enforced as concepts for consistency and repeatability in verification procedures (inspection).

Introduction to Geometric Dimensioning and Tolerancing
Possible results of several different manufactured parts are examined based upon various dimensioning and tolerancing techniques. The need for greater control of tolerancing specifications is established, and introductory materials are presented to demonstrate current solutions, based upon the national and international standards (ASME Y14.5 and ISO 9000). The unique symbolism associated with these tolerancing standards is reviewed and illustrated. Control limitations associated with the fundamental rules of geometric dimensioning and tolerancing are also discussed, and specific form tolerances (straightness of a feature, and straightness of a feature of size) are introduced and illustrated.

Tolerances of Form
Straightness of features and features of size is continued, and the differences between the two straightness applications are examined. The impact of geometric controls imposed on features of size is emphasized. Feature control frames and the organization of information they contain is reviewed and clarified—specifically in applications controlling axis straightness of cylindrical components. Flatness control is introduced and examples illustrated. Explanations of flatness control and its associated limitations under the specifics of size dimensions and the first fundamental rule are examined, and differences between straightness and flatness controls are characterized.

Circularity of features, and feature cylindricity are discussed. Both of these controls are explained and their differences clarified. Examples of circular and cylindrical parts are illustrated, and the extent of specific controls explained. Comparisons between the actual types of control between these specifications and straightness and flatness are provided, and emphasis is given to the uniqueness of the straightness form control when applied to a feature of size. Fundamental rule number one (the envelope rule, or Taylor Principle) is examined as an automatic form control based upon size dimension limitations, and the fact that it is overridden when a geometric tolerance is applied to a feature of size. The basic principle is stressed that straightness, when applied to a feature of size, is the only form control that may be applied to features of size, but that all form tolerances can be applied to features.

Datums and Datum References
Datums are defined as theoretically perfect points, lines, axes, and/or planes. Explanations are given on how datums are used within the system of geometric controls and measurements. Illustrations are provided which explain how datums are established in the manufacturing and inspection phases of development. Theoretical datums, datum simulators, and datum features are explained and the differences between them considered. The students are introduced to the concept of a datum reference system and explanations are given about how such datum reference systems are used in the manufacture and verification processes.

Datum feature selection and datum referencing are also considered. Specific geometric characteristics and functional relationships are discussed and illustrated as crucial elements in the datum selection process. Primary, secondary and tertiary datums are explained along with their functional relationships to the specified geometric control. Datum precedence is introduced as a crucial element in datum referencing, and the associated order and symbolism explained. Several examples of datum referenced geometric controls are examined. The procedure used to establish a feature axis or a feature center plane as a datum is also explained.

are carefully examined and illustrated. Specifications for total runout tolerances are reviewed as they relate to their application on engineering drawings.

Geometric Tolerances—A General Review

A review of geometric tolerancing philosophies and results are given and compared to what may be expected when independent coordinate tolerancing is used. Major advantages offered by this system of controls are considered, as well as drawbacks or disadvantages resulting from inappropriate or misinformed applications. Discussions include cost and time to market considerations, when this system of controls is appropriate for use, and common educational and industrial concerns about teaching and using geometric tolerancing as a system of tolerance controls.

Practice and Self-Evaluation Problems

To avoid confusion and to allow the pages to be removed, the practice and self-evaluation problems are printed on only one side of each page. The blank sheets can be used for notes or for sketching purposes. Solutions for the practice problems are included on the CD-ROM. You are encouraged to attempt the problems prior to reviewing the respective solution.

For each unit of study, there is a series of *practice problems* in this book that are coordinated with presentations on the CD-ROM. Each of the practice problems that are in the book are provided a solution on the CD-ROM.

You are strongly encouraged to complete each of the practice problems prior to viewing the solutions.

In addition to the practice problems, the first four units contain specific *self-evaluation problems*. There are no solution slides provided on the CD-ROM for the self-evaluation problems. The self-evaluation problems are labeled, and instructions for completing them are included, both in the presentations, and on the work sheets themselves.

Orthographic Projection
This section contains 9 practice problems (problems 1-9) following by a self-evaluation consisting of 4 problems (problems 10-14).

Auxiliary Views
This section contains 4 practice problems (problems 15-18) following by a self-evaluation consisting of 4 problems (problems 19-22).

Sectioning Conventions
This section contains 7 practice problems (problems 23-29) following by a self-evaluation consisting of 6 problems (problems 30-35).

Coordinate Dimensioning and Tolerancing
This section contains 4 practice problems (problems 36-39) following by a self-evaluation consisting of 5 problems (problems 40-44).

Introduction to Geometric Tolerancing
This section contains 4 quizzes (problems 45-48).

Tolerances of Form
This section contains 9 practice problems and quizzes (problems 49-57).

Datums and Datum Targets
This section contains 7 practice problems and quizzes (problems 58-64).

Orientation Tolerances
This section contains 6 practice problems and quizzes (problems 65-70).

Tolerances of Location
This section contains 7 practice problems and quizzes (problems 71-77).

Tolerances of Profile
This section contains 7 practice problems and quizzes (problems 78-84).

Tolerances of Runout
This section contains 4 practice problems and quizzes (problems 85-88).

Orthographic Projection Problems

The following problems correspond with the slides in the *Introduction* and *Orthographic Projection* presentation. Follow the instructions for each problem, and for all but the special slides labeled *"Self Evaluation"*, solution slides are available in the presentation materials on the CD.

Problem #1—Orthographic Projection

Complete the views below by projecting all missing lines.

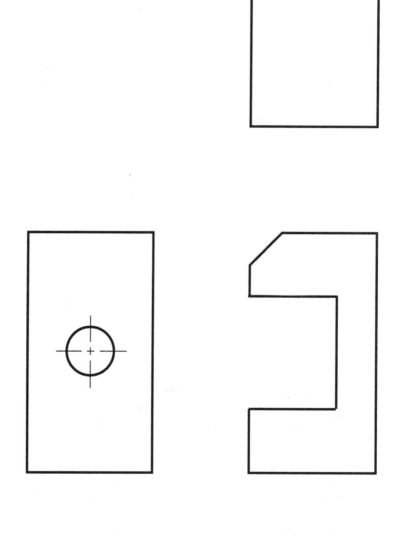

Problem #2—Orthographic Projection

Complete the views below by adding any missing lines in all three views.

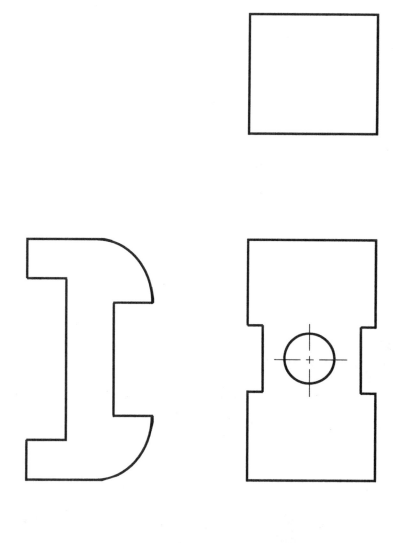

Problem #3—Orthographic Projection

Complete the views below by adding any missing lines in all three views.

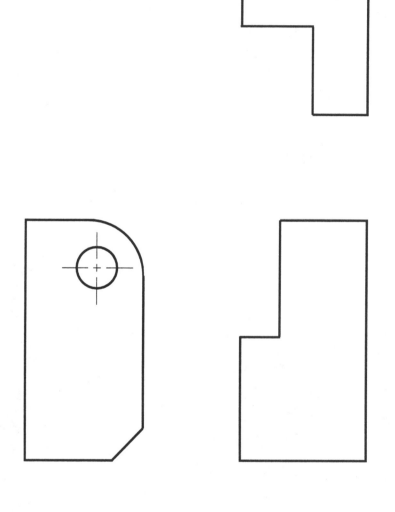

Problem #4—Orthographic Projection

Complete the views below by adding any missing lines in all three views.

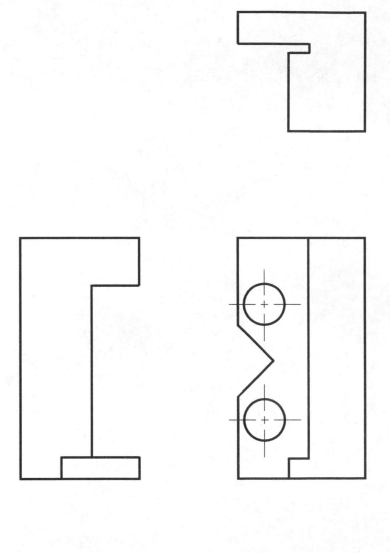

Problem #3—Orthographic Projection

Complete the views below by adding any missing lines in all three views.

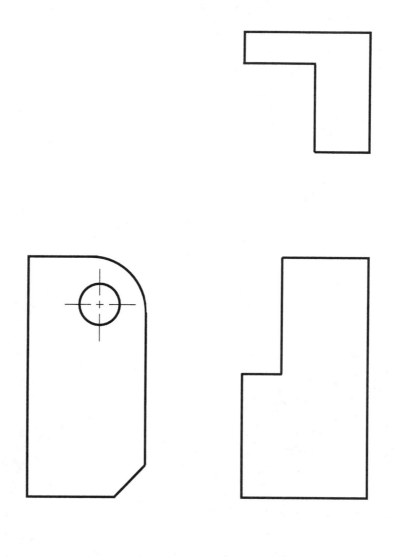

Problem #4—Orthographic Projection

Complete the views below by adding any missing lines in all three views.

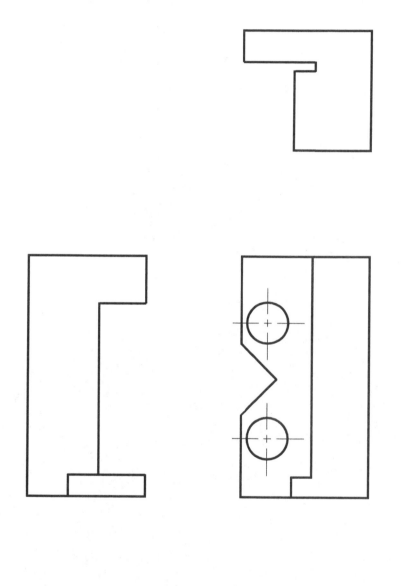

Problem #5—Orthographic Projection

Complete the views below by adding any missing lines in all three views.

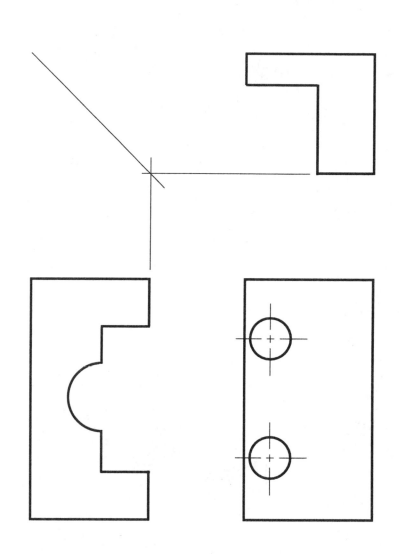

Problem #6—Orthographic Projection

Complete the views below by adding any missing lines in all three views.

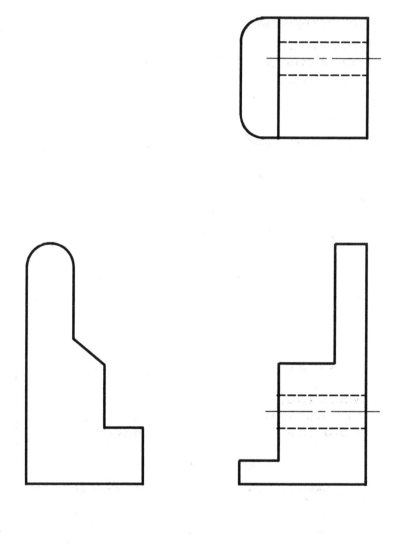

Problem #7—Orthographic Projection

Complete the views below by adding any missing lines in all three views.

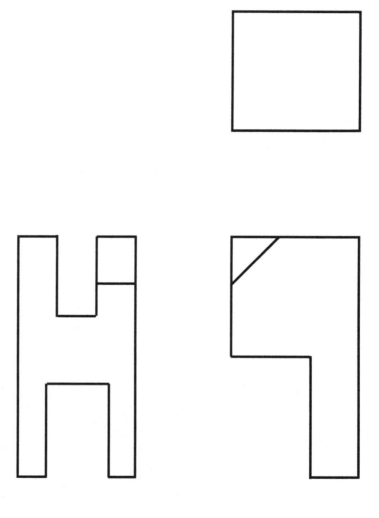

Problem #8—Orthographic Projection

Using the principles of orthographic projection, construct the front, top, and right side views of the object shown below.

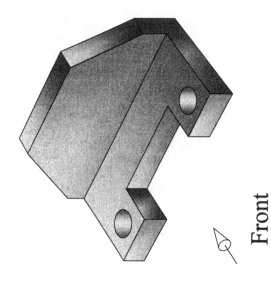

Front

Problem #9—Orthographic Projection

Using the principles of orthographic projection, construct the front, top, and right side views of the object shown below.

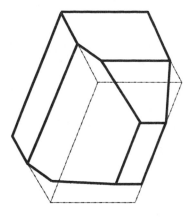

Orthographic Projection
SELF EVALUATION

There are four self-evaluation problems in this section.

Problem #10—Orthographic Projection

Self Evaluation (Consisting of Problems 10 – 14)

Complete all three views by adding any lines that may be missing. Make certain that all object, hidden, and center lines are accounted for in each of the three views.

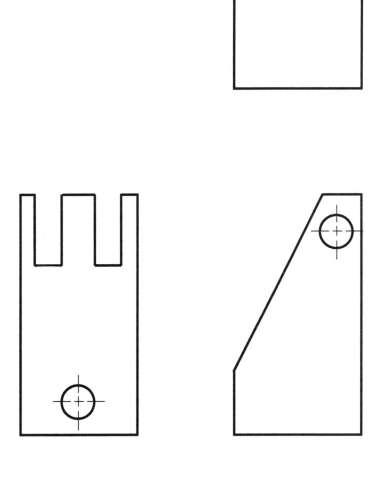

Problem #11—Orthographic Projection
Self Evaluation (Consisting of Problems 10 – 14)

Complete all three views by adding any lines that may be missing. Make certain that all object, hidden, and center lines are accounted for in each of the three views.

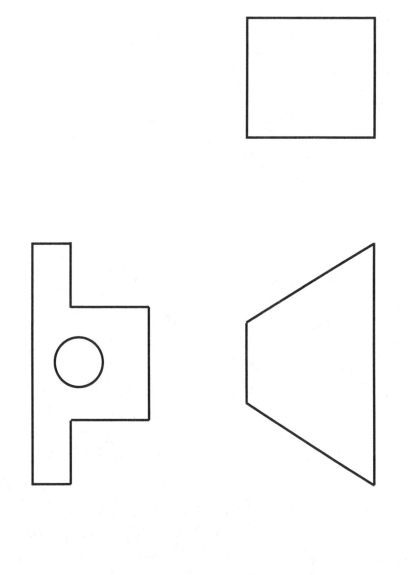

Problem #12—Orthographic Projection
Self Evaluation (Consisting of Problems 10 – 14)

Complete all three views by adding any lines that may be missing. Make certain that all object, hidden, and center lines are accounted for in each of the three views. *Problems 10-14 in your packet are to be turned in for evaluation and grading* (see due date for assignment #1 on your class schedule).

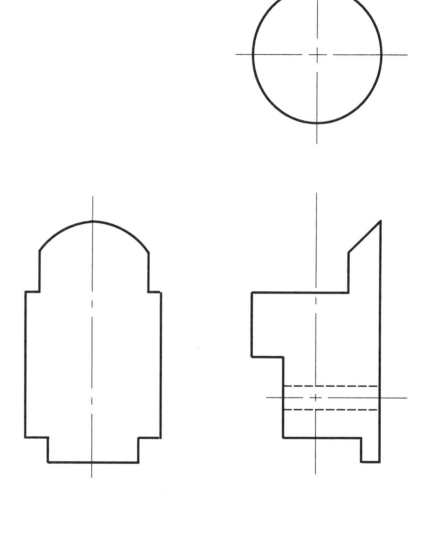

Problem #14—Orthographic Projection
Self Evaluation (Consisting of Problems 10 – 14)

Complete all three views by adding any lines that may be missing. Make certain that all object, hidden, and center lines are accounted for in each of the three views (the small hole in the front view is a threaded (tapped) hole). *Problems 10-14 in your packet are to be turned in for evaluation and grading* (see due date for assignment #1 on your class schedule).

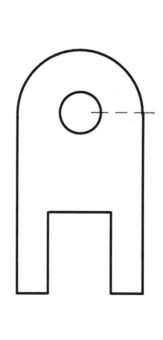

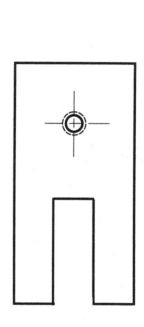

Auxiliary Views Projection Problems

The following problems correspond with the slides in the *Auxiliary Views* presentation. Follow the instructions for each problem, and for all but the special slides labeled *"Self Evaluation"*, solution slides are available in the presentation materials on the CD.

Problem #15—Auxiliary Views

Oblique Surfaces—True Length Line Given

Solve for the true size and shape of the surface.

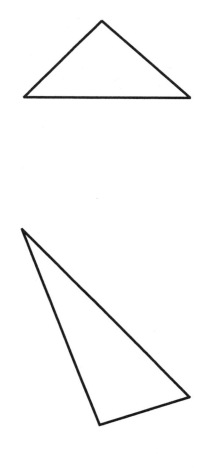

Problem #16—Auxiliary Views
True-Angle Problem

Solve for the true angle between side A and side B. Use the view on the right as the point of departure for projecting the primary auxiliary view.

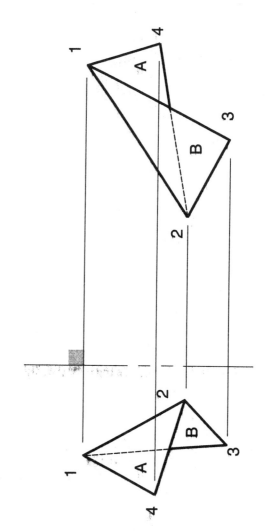

Problem #17—Auxiliary Views
Slope Angle of Oblique Lines

Construct the required auxiliary view(s) that will display the true slope angle.

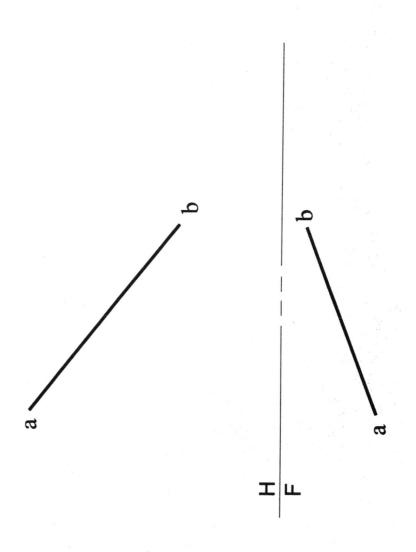

Problem #18—Distance From a Point to a Plane

Solve for the shortest (perpendicular), vertical, and horizontal distances from point A to the plane defined by points 1, 2, and 3. Also solve for the slope angle of plane 1, 2, 3.

Horizontal View

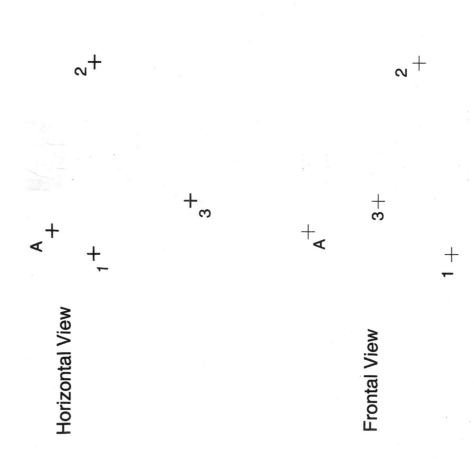

Frontal View

Auxiliary Views
SELF EVALUATION

There are four self-evaluation problems in this section.

Problem #19—Auxiliary Views
Self Evaluation (Consisting of Problems 19-22)

Project a view that displays the true size and true shape of surface A. The two given views are complete as shown (and labeled correctly).

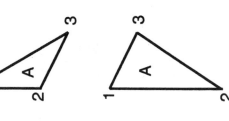
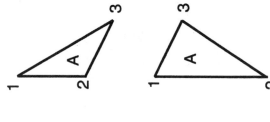

Problem #20—Auxiliary Views
Self Evaluation (Consisting of Problems 19-22)

From the two views of the pyramid, construct true size and shape views of side A and side C. Also, construct a view showing the true angle between the two sides (A and C).

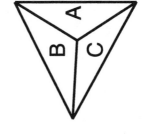

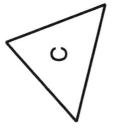

Problem #21—Auxiliary Views
Self Evaluation (Consisting of Problems 19-22)

Construct a true size and shape view of surface B and include the entire object in your answer. Remember, surface B must be closest to your eye in the true size and shape view.

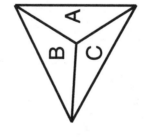

Problem #22—Auxiliary Views

Self Evaluation (Consisting of Problems 19-22)

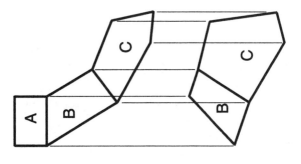

Sectioning Problems

The following problems correspond with the slides in the *Sectioning and Sectioning Conventions* presentation. Follow the instructions for each problem, and for all but the special slides labeled "*Self Evaluation*", solution slides are available in the presentation materials on the CD.

Problem #23—Sectioning Practice Problems

Construct the indicated section view. Then, when completed, advance to the solution slide in the presentation, to check your work.

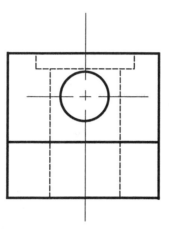

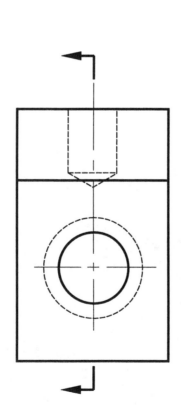

Problem #24—Sectioning Practice Problems

Construct the indicated section view. Then move on to the solution slide which is the next slide in the presentation.

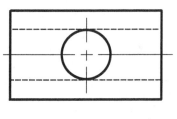

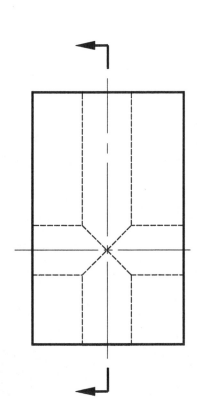

Problem #25—Sectioning Practice Problems

Construct the required section view. When you have completed the problem, advance to the solution slide, which is the next slide in the presentation.

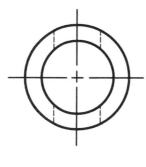

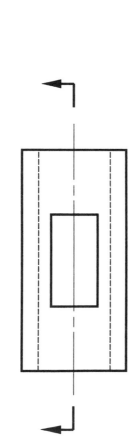

Problem #27—Sectioning Practice Problems

Construct the right side view as the appropriate section view that would result from having the cutting plane positioned as shown in the front view. When complete, advance to the solution slide.

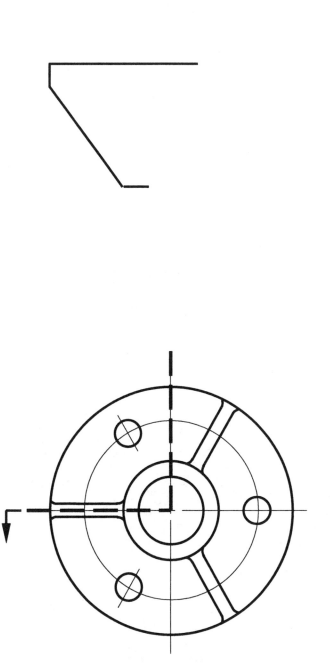

Problem #28—Sectioning Practice Problems

Convert the right side view into a correct full section view.

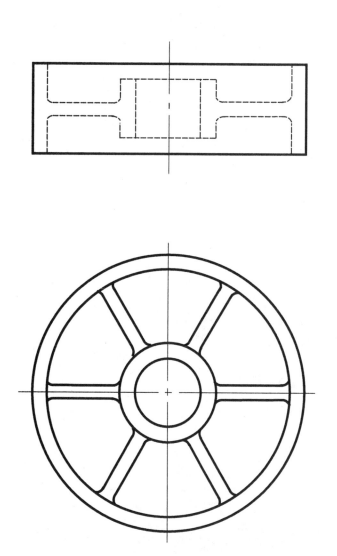

Problem #29—Sectioning Practice Problems

Convert the right side view into a full section view. The webs in this case, are the same depth (in the right side view) as the outside rim of the casting. Use the appropriate crosshatch pattern.

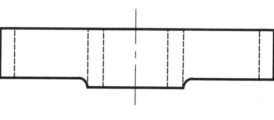

Sectioning
SELF EVALUATION

There are six self-evaluation problems in this section.

Problem #30—Sectioning Conventions
Self Evaluation—(Problems 30-35)

Complete the front view as the indicated full section view.

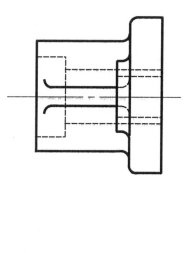

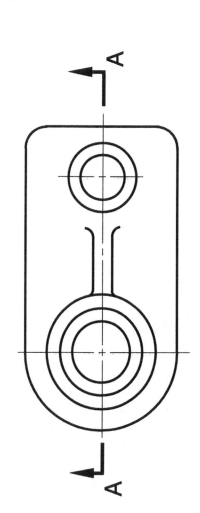

SECTION A – A

Problem #31—Sectioning Conventions
Self Evaluation—(Problems 30-35)

Convert the front view into the required section view.

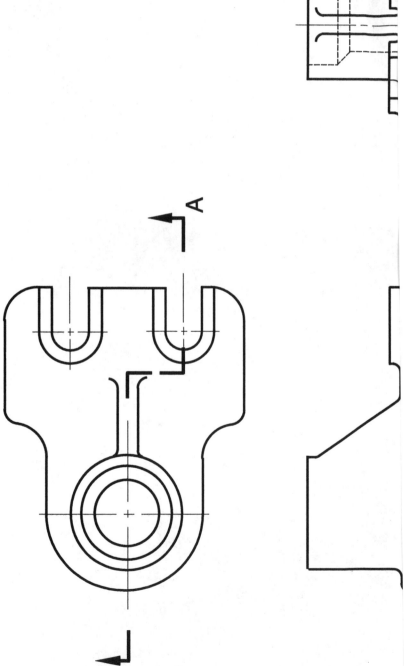

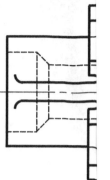

Problem #32—Sectioning Conventions
Self Evaluation—(Problems 30-35)

Construct a full section in the top view, as indicated.

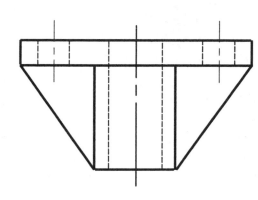

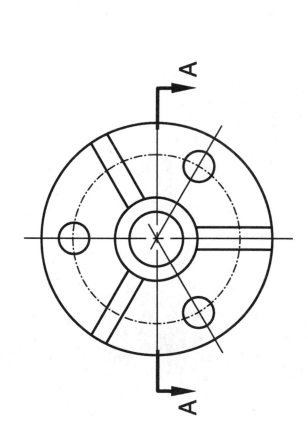

SECTION A - A

A

A

Problem #33—Sectioning Conventions
Self Evaluation—(Problems 30-35)

Construct a half section in the right side view, as indicated.

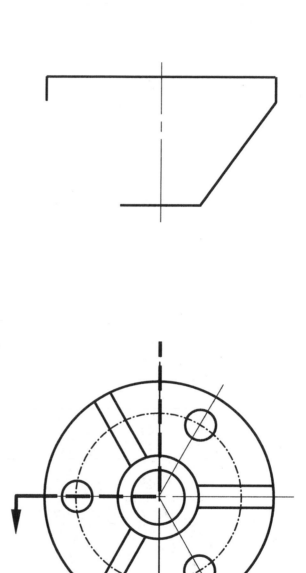

Problem #34—Sectioning Conventions
Self Evaluation—(Problems 30-35)

Re-draw the front view of the object as a full section view, in the indicated space.

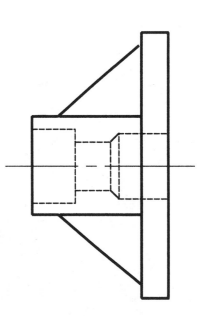

SECTION C - C

Problem #35—Sectioning Conventions
Self Evaluation—(Problems 30-35)

Complete the required section view.

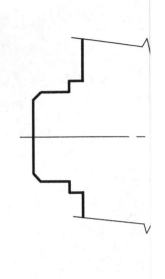

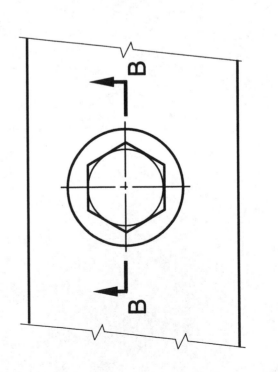

B

B

Coordinate Dimensioning and Tolerancing Problems

The following problems correspond with the slides in the *Coordinate Dimensioning and Tolerancing* presentation. Follow the instructions for each problem. For several of these problems there are solution slides provided in the presentation. However, for the problems included in *"Self Evaluatoin"*, no solution slides have been made available in the presentation materials on the CD. Carefully review the content of the slide presentation (Lesson 4) before completing these problems.

Problem #37—Dimensioning and Tolerancing
Material Condition and Dimension Variation

Fill in the chart below, indicating the maximum and least material condition values resulting from possible variations.

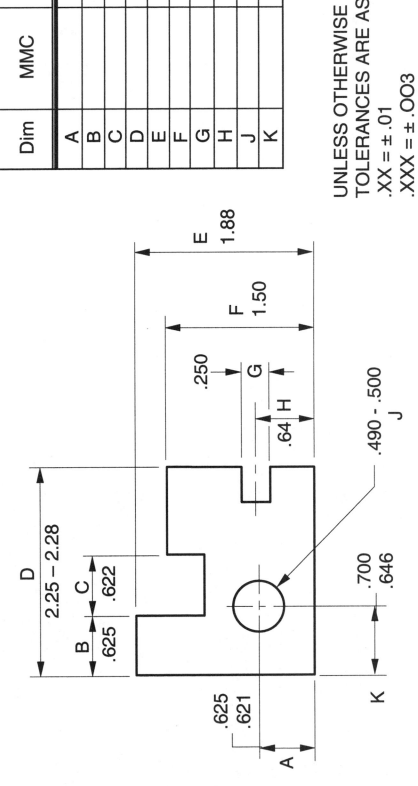

Dim	MMC	LMC
A		
B		
C		
D		
E		
F		
G		
H		
J		
K		

UNLESS OTHERWISE SPECIFIED,
TOLERANCES ARE AS FOLLOWS:
.XX = ±.01
.XXX = ±.003

Problem #38—Dimensioning and Tolerancing
Basic Hole System

Using the basic hole system, apply the dimensions on the drawing. Use stacked limits when placing the information on the drawing.

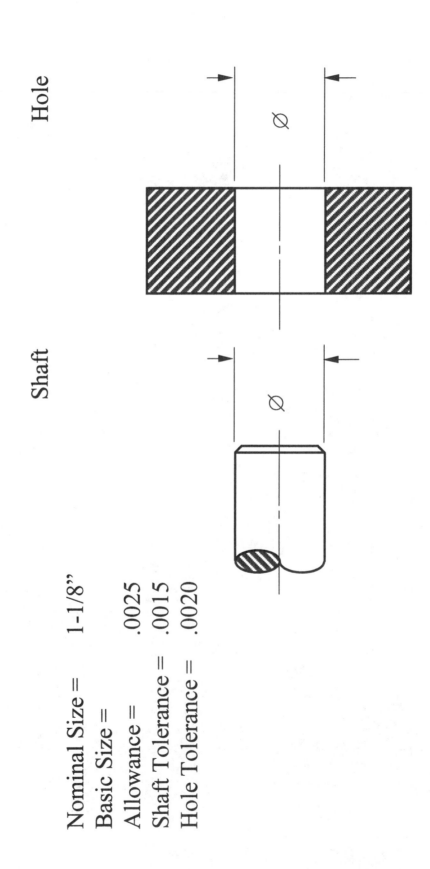

Shaft

Hole

Nominal Size =   1-1/8"

Basic Size =

Allowance =   .0025

Shaft Tolerance =   .0015

Hole Tolerance =   .0020

Problem #39—Dimensioning and Tolerancing
Basic Shaft System

Using the *Basic Shaft System* and the following data, complete the drawing by placing the appropriate dimensions on both the shaft and the hole. Use the *stacked limits* expression for the dimensions.

Nominal Size =   9/16"

Basic Size =

Allowance =   .0004

Shaft Tolerance =   .0008

Hole Tolerance =   .0010

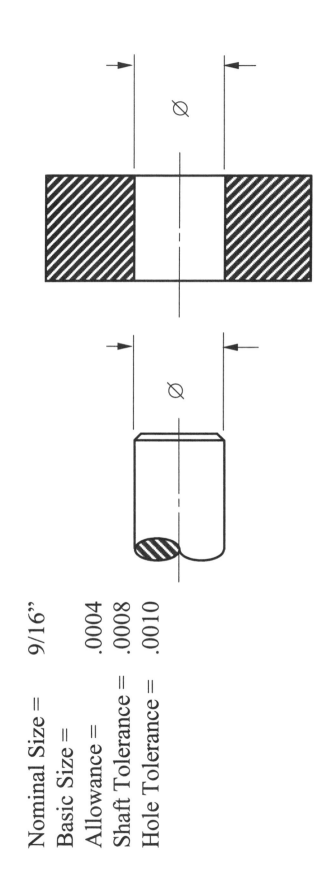

Problem #40—Dimensioning and Tolerancing
Self Evaluation—Using the Standard Fit Tables

Place upper and lower limits dimensions for the shaft and hole in the drawing below. Specify a *precision slow-running, low journal-pressure* fit. The nominal diameter of the shaft and hole is 7/8". Refer to next page in this book for fit data.

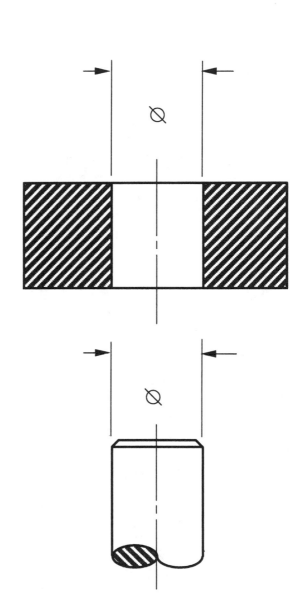

Running and Sliding Fits* -- American National Standard

RC 1   Close sliding fits are intended for the accurate location of parts which must assemble without perceptible play.

RC 2   Sliding fits are intended for accurate location, but with greater maximum clearance than class RC 1. Parts made to this fit move and turn easily but are not intended to run freely, and in the larger sizes may seize with small temperature changes.

RC 3   Precision running fits are about the closest fits which can be expected to run freely, and are intended for precision work at slow speeds and light journal pressures, but are not suitable where appreciable temperature differences are likely to be encountered.

RC 4   Close running fits are intended chiefly for running fits on accurate machinery with moderate surface speeds and journal pressures, where accurate location and minimum play are desired.

RC 5
RC 6   Medium running fits are intended for higher running speeds, or heavy journal pressures, or both.

Basic hole system. Limits are in thousandths of an inch.

Nominal Size Range in Inches	Class RC 1			Class RC 2			Class RC 3			Class RC 4			Class RC 5			Class RC 6		
	Limits of Clearance	Standard Limits Hole H5	Shaft g4	Limits of Clearance	Standard Limits Hole H6	Shaft g5	Limits of Clearance	Standard Limits Hole H7	Shaft f6	Limits of Clearance	Standard Limits Hole H8	Shaft f7	Limits of Clearance	Standard Limits Hole H8	Shaft e7	Limits of Clearance	Standard Limits Hole H9	Shaft e8
0 - 0.12	0.1 / 0.45	+0.2 / -0	-0.1 / -0.25	0.1 / 0.55	+0.25 / -0	-0.1 / -0.3	0.3 / 0.95	+0.4 / -0	-0.3 / -0.55	0.3 / 1.3	+0.6 / -0	-0.3 / -0.7	0.6 / 1.6	+0.6 / -0	-0.6 / -1.0	0.6 / 2.2	+1.0 / -0	-0.6 / -1.2
0.12 - 0.24	0.15 / 0.5	+0.2 / -0	-0.15 / -0.3	0.15 / 0.65	+0.3 / -0	-0.15 / -0.35	0.4 / 1.12	+0.5 / -0	-0.4 / -0.7	0.4 / 1.6	+0.7 / -0	-0.4 / -0.9	0.8 / 2.0	+0.7 / -0	-0.8 / -1.3	0.8 / 2.7	+1.2 / -0	-0.8 / -1.5
0.24 - 0.40	0.2 / 0.6	+0.25 / -0	-0.2 / -0.35	0.2 / 0.85	+0.4 / -0	-0.2 / -0.45	0.5 / 1.5	+0.6 / -0	-0.5 / -0.9	0.5 / 2.0	+0.9 / -0	-0.5 / -1.1	1.0 / 2.5	+0.9 / -0	-1.0 / -1.6	1.0 / 3.3	+1.4 / -0	-1.0 / -1.9
0.40 - 0.71	0.25 / 0.75	+0.3 / -0	-0.25 / -0.45	0.25 / 0.95	+0.4 / -0	-0.25 / -0.55	0.6 / 1.7	+0.7 / -0	-0.6 / -1.0	0.6 / 2.3	+1.0 / -0	-0.6 / -1.3	1.2 / 2.9	+1.0 / -0	-1.2 / -1.9	1.2 / 3.8	+1.6 / -0	-1.2 / -2.2
0.71 - 1.19	0.3 / 0.95	+0.4 / -0	-0.3 / -0.55	0.3 / 1.2	+0.5 / -0	-0.3 / -0.7	0.8 / 2.1	+0.8 / -0	-0.8 / -1.3	0.8 / 2.8	+1.2 / -0	-0.8 / -1.6	1.6 / 3.6	+1.2 / -0	-1.6 / -2.4	1.6 / 4.8	+2.0 / -0	-1.6 / -2.8
1.19 - 1.97																		
1.97 - 3.15																		

*From ANSI B4.1--1967 (R 1987). Larger diameters and RC 7 through RC 9 not included in this presentation.

Problem #41—Dimensioning and Tolerancing
Self Evaluation—Dimensioning Screw Threads

Referencing the data on the next page (65), determine the necessary information and define the internal and external threads in the drawings below. (The major diameter is ¾ inch, the threads are a class 2 fit, their form and series is Unified National Fine).

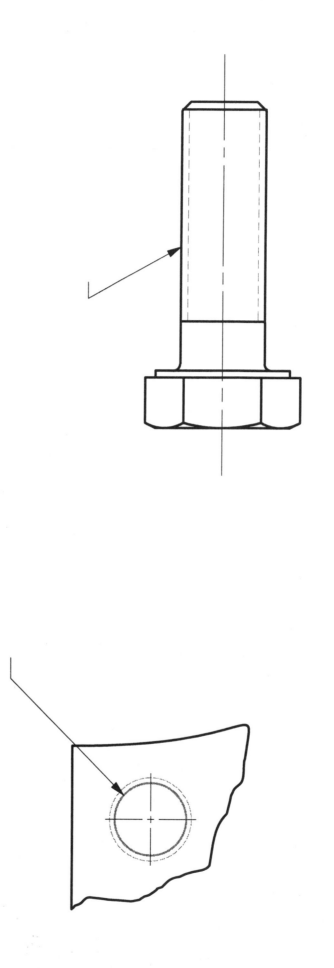

American National Standard, Unified, and American National Screw Threads (Partial Table)

Nominal Diameter	Coarse[a] NC and UNC Thds. per inch	Fine[a] NF and UNF Thds per inch	Extra Fine[b] NEF and UNEF Thds per inch
0 (.060)	...	80	...
1 (.073)	64	72	...
2 (.086)	56	64	...
3 (.099)	48	56	...
4 (.112)	40	48	...
5 (.125)	40	44	...
6 (.138)	32	40	...
8 (.164)	32	36	...
10 (.190)	24	32	...
12 (.216)	24	28	32
1/4	20	28	32
5/16	18	24	32
3/8	16	24	32
7/16	14	20	28
1/2	13	20	28
9/16	12	18	24
5/8	11	18	24
11/16	...	...	24
3/4	10	16	20
13/16	...	...	20
7/8	9	14	20
15/16	...	...	20
1	8	12	20

[a]Classes 1A, 2A, 3A, 1B, 2B, 3B, 2, and 3

[b]Classes 2A, 2B, 2, and 3

Problem #42—Dimensioning and Tolerancing

Self Evaluation—Dimensioning Exercise

Dimension the *internal* 7-3/4 ±1/32" diameter on the drawing below. Use an *"arrows in, dimension in"* technique. Also, locate and dimension the four mounting holes on a bolt circle of 9 inches—plus or minus 1/16", and which have a major diameter of 11/16" (the mounting hole tolerance is twenty thousandths). Use *decimal* limits dimensioning throughout.

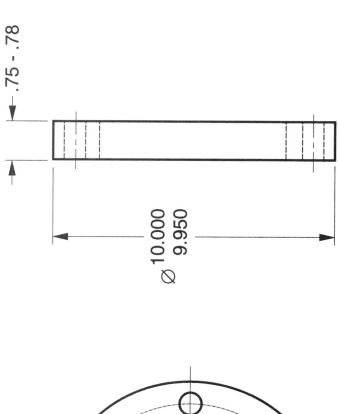

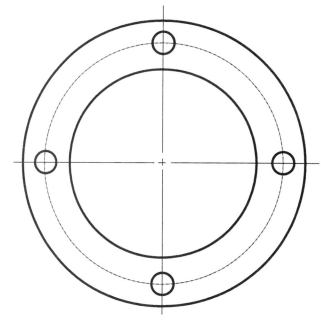

Problem #44—Dimensioning and Tolerancing

Self Evaluation—Dimensioning Exercise

Without specifying actual dimension *values*, show all dimensions necessary to locate the holes in the part below, using *coordinate* arrangement rather than polar oriented dimensions.

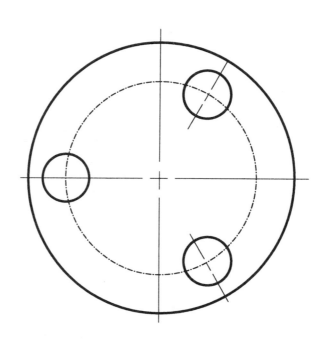

Introduction to Geometric Tolerancing

Problems

The following problems correspond with the slides in the *Introduction to Geometric Dimensioning and Tolerancing* presentation. Follow the instructions for each problem. For all of these problems there are solution slides provided in the presentation. The concepts are very important and will be called upon many times in future activities.

Problem #45—GD&T Introduction

Complete this quiz as follows: On the line before the symbol, identify the *geometric characteristic symbol*. Following the symbol, indicate whether it is an *individual* or *related* control.

———— ⊥   Individual ⊠ Related ⊠

———— ⟋   Individual ⊠ Related ⊠

———— ⚡   Individual ⊠ Related ⊠

———— ⌀⟋ Individual ⊠ Related ⊠

———— ⌓   Individual ⊠ Related ⊠

———— ◎   Individual ⊠ Related ⊠

———— ⟋   Individual ⊠ Related ⊠

———— —   Individual ⊠ Related ⊠

———— ⌒   Individual ⊠ Related ⊠

———— ○   Individual ⊠ Related ⊠

———— ∠   Individual ⊠ Related ⊠

———— ☰   Individual ⊠ Related ⊠

———— ⊕   Individual ⊠ Related ⊠

———— ⫽   Individual ⊠ Related ⊠

Problem #47—GD&T Introduction

1. Datum references are placed between the tolerance zone descriptor and the tolerance value, inside the feature control frame.   True ☒   False ☒

2. Feature control frames have a required format to be followed by users of GD&T.   True ☒   False ☒

3. A dimension value inside brackets is considered a basic dimension.   True ☒   False ☒

4. The first fundamental rule of geometric dimensioning and tolerancing cannot be overridden or superseded.   True ☒   False ☒

5. The first fundamental rule (Rule #1) applies to all dimensions on the drawing.   True ☒   False ☒

6. There are no exceptions to the first fundamental rule of GD&T.   True ☒   False ☒

7. The second fundamental rule of GD&T describes conventions for expressing geometric tolerances in feature control frames.   True ☒   False ☒

8. There are _____ *geometric characteristic symbols* used in GD&T.

9. Feature control frames may be used in conjunction with control of features and/or features of size.   True ☒   False ☒

Tolerances of Form
Problems

The following problems correspond with the slides in the *Tolerances of Form* presentation. Follow the instructions for each problem. For all of these problems there are solution slides provided in the presentation.

Problem #49—Feature Straightness
Application Quiz

1. The geometric tolerance control for feature straightness must always be applied to a surface line element.

   True ☐   False ☐

2. The straightness control must always be applied in a view where the surface elements are shown as a straight line.

   True ☐   False ☐

3. Modifiers may be used in the feature control frame for accuracy when controlling feature straightness.

   True ☐   False ☐

4. The straightness control should always be a refinement of other geometric tolerances which indirectly affect the feature straightness.

   True ☐   False ☐

5. In certain situations, datums may be used in the feature control frame.

   True ☐   False ☐

Problem #50—Straightness Applied to a Feature of Size
Application Quiz

1. For axis control, the geometric tolerance may be applied to either a line element or a feature of size.

   True ☐   False ☐

2. The straightness geometric control must be associated with a size dimension.

   True ☐   False ☐

3. Modifiers must be used in the feature control frame when controlling size-feature straightness.

   True ☐   False ☐

4. The straightness control should be a refinement of other geometric tolerances which indirectly affect the feature straightness.

   True ☐   False ☐

5. Datums may not be used in the feature control frame.

   True ☐   False ☐

Problem #52—Circularity Applied to a Feature Application Quiz

1. For circularity control, the geometric tolerance may be applied to either a feature or feature of size.

   True ☐   False ☐

2. Circularity control refers to its own perfect geometric counterpart; always applied to a diametrical feature.

   True ☐   False ☐

3. Modifiers may be used in the feature control frame when controlling circularity.

   True ☐   False ☐

4. The circularity control must be a refinement of other geometric tolerances which indirectly affect the circularity characteristics of the feature.

   True ☐   False ☐

5. Datum references may not be used with circularity.

   True ☐   False ☐

Problem #53—Cylindricity Applied to a Feature Application Quiz

1. For cylindricity control, the geometric tolerance must be applied to a feature of size.

   True ☐   False ☐

2. Cylindricity control can override the effects of Rule #1 because it is applied to a feature of size.

   True ☐   False ☐

3. Modifiers may not be used in the feature control frame when controlling cylindricity.

   True ☐   False ☐

4. The cylindricity control need not be a refinement of other geometric tolerances which indirectly affect the cylindricity characteristics of the feature.

   True ☐   False ☐

5. Datum references may not be used with cylindricity.

   True ☐   False ☐

Problem #54—Form Tolerance Applications

Ø .495 ± .005

| .xxx |

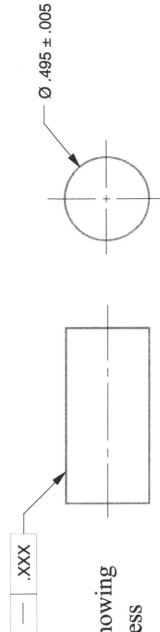

Complete the table by showing the total out of straightness tolerance that would be permitted for the specified *actual* sizes.

Allowable Straightness Error

Actual Size Ø		.004
.500		
.492		
.501		
.498		
.490		
.495		

Problem #56—Form Tolerance Applications

Respond to the questions below.

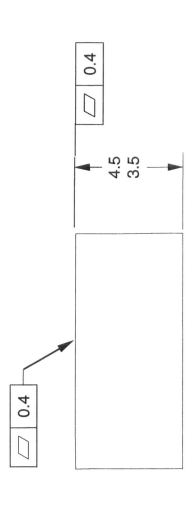

The alternative placement of the feature control frame would produce the same result.

True ☐ False ☐

What is the allowable flatness error of the top surface when the height is at MMC?

_____

What is the allowable flatness error of the top surface when the height is at LMC?

_____

The flatness tolerance shown in the figure above applies RFS.

True ☐ False ☐

Datums and Datum Targets
Problems

The following problems correspond with the slides in the *Datums and Datum Targets* presentation. Follow the instructions for each problem. For all of these problems there are solution slides provided in the presentation.

Problem #59— Datums and Datum Targets

Complete the following problems (58-60) before advancing to the solution slides in the main presentation.

5. When placed in contact with a surface plate or machine table, a primary flat surface will always stabilize on a minimum of _____ points.

    a. ☒ one

    b. ☒ two

    c. ☒ three

    d. ☒ all of the above

6. A datum feature of size is identified on the drawing by placing the datum feature symbol _____.

    a. ☒ on the feature's extension line if no space is available on its center line

    b. ☒ anywhere near the feature of size

    c. ☒ on the dimension, or on an extension of the dimension line

    d. ☒ on the feature object line

7. To establish a datum axis, the datum feature symbol must be placed on the representative center line of the feature   ☒ True   ☒ False

8. Datum precedence, as delineated in the feature control frame, affects how datum features on the part are used to establish the datum reference frame. True ☒   False ☒

Page 165

Problem #60— Datums and Datum Targets

Complete the following problems (58-60) before advancing to the solution slides in the main presentation.

9. The minimum number of points required to establish a *flat surface* as a secondary datum plane is

   a. ☒ one

   b. ☒ two

   c. ☒ three

   d. ☒ All of the above.

10. Because datums are theoretical entities that do not actually exist on real parts, they must be simulated by the use of gages and other inspection tools.

    True ☒   False ☒

11. It is important to maintain alphabetical order in referencing datums in feature control frames.   ☒ True   ☒ False

12. If tolerance specifications reference datums, verifying measurements must be taken from the datums rather than from the associated datum features.

    True ☒   False ☒

13. Datum features that are referenced as primary in one specification can not be referenced as secondary or tertiary in other feature control frames.   True ☒

    False ☒

Problem #61—Datums and Datum Targets

In either or both views, identify (1) a datum feature, (2) a datum feature symbol, (3) a datum reference, (4) a simulated datum, and (5) datum A.

DRAWING

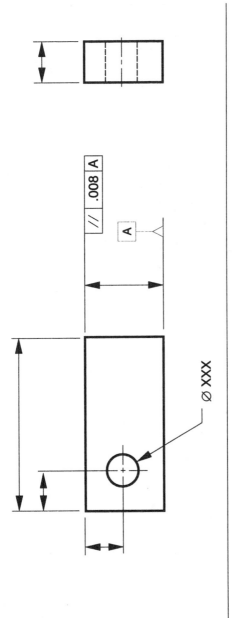

⌀ xxx

INTERPRETATION

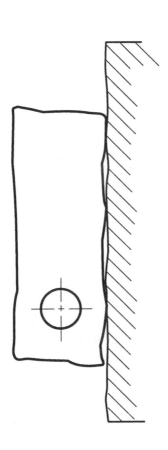

Problem #62—Datums and Datum Targets

Explain how datum A would be established as a primary datum, *regardless of feature size*, in each of the problems shown below.

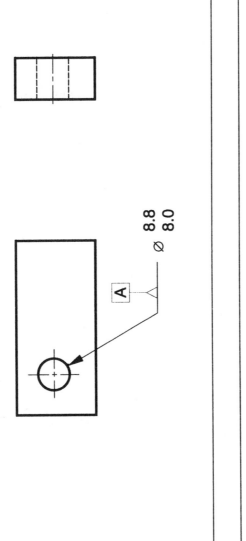

Ø 8.8
8.0

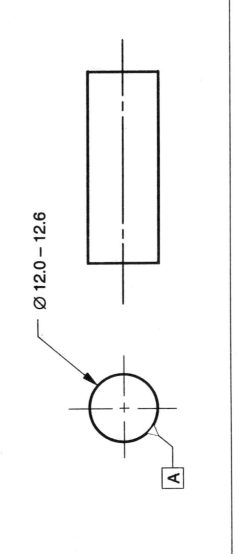

Ø 12.0 – 12.6

Problem #63—Datums and Datum Targets

Explain how datum A would be established as a primary datum, *regardless of feature size?*

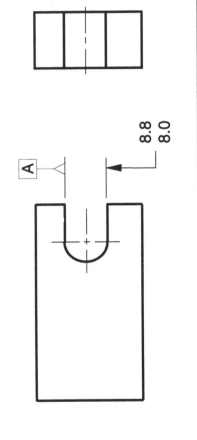

8.8
8.0

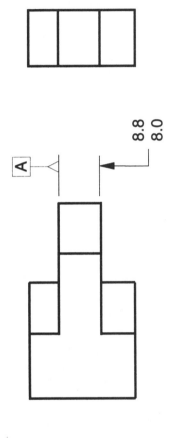

8.8
8.0

Problem #64—Datums and Datum Targets

The primary datum is A, datum B is secondary, datum C is tertiary. (1) correctly label all datum features, and datum targets (2) datum target location dimensions are _____, (3) Datums C and B are established by datum target _____ [points, lines, areas], (4) Datum A is on the _____ surface of the part.

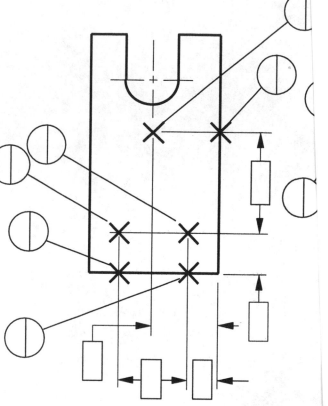

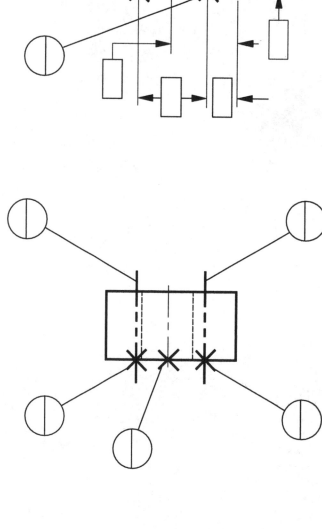

Problem #63—Datums and Datum Targets

Explain how datum A would be established as a primary datum, *regardless of feature size?*

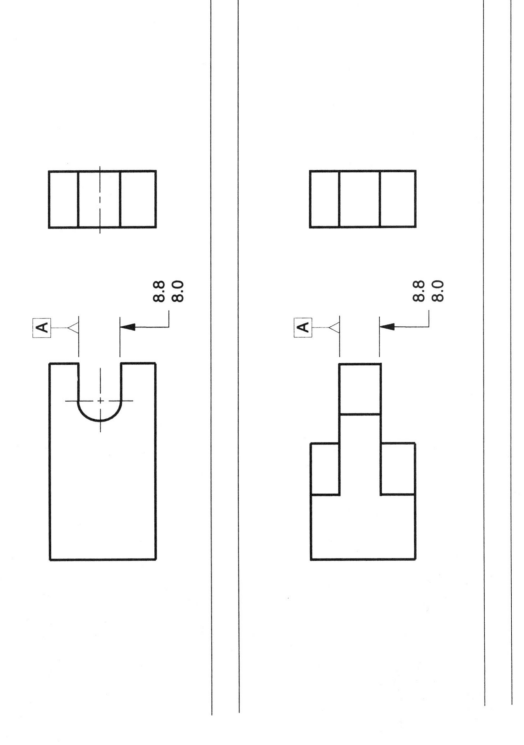

Problem #64—Datums and Datum Targets

The primary datum is A, datum B is secondary, datum C is tertiary. (1) correctly label all datum features, and datum targets (2) datum target location dimensions are _____, (3) Datums C and B are established by datum target _____ [points, lines, areas], (4) Datum A is on the _____ surface of the part.

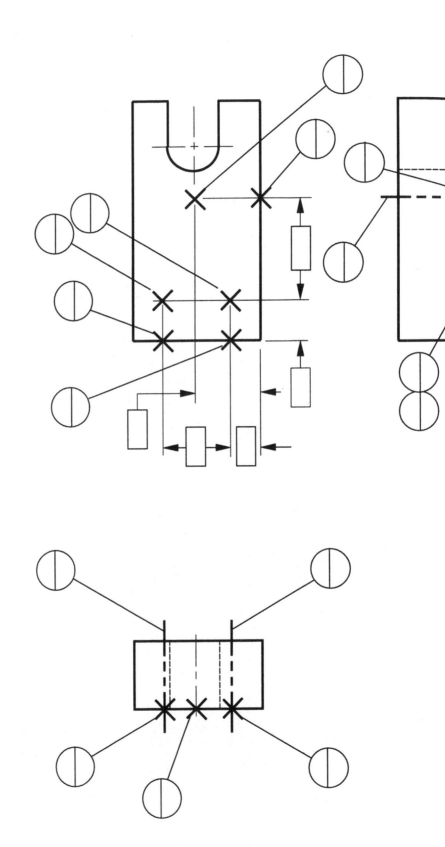

Orientation Tolerances
Problems

The following problems correspond with the slides in the *Tolerances of Orientation* presentation. Follow the instructions for each problem. For all of these problems there are solution slides provided in the presentation.

Problem #66 — Parallelism

Complete before advancing to the solution slide.

1. True ☒ False ☒   When applied to one side of a rectangular object, parallelism controls the physical size limits as well as the relationship between the feature being controlled and the datum feature.

2. Parallelism, when applied to _____, may actually result in surface errors that fall outside the boundaries of the parallelism tolerance zone, but a plane contacting the high points of the surface will be within the tolerance zone.

   a. ☒ an individual feature

   b. ☒ a tangent plane

   c. ☒ a unit area (defined by chain lines and located with basic dimensions)

   d. ☒ a pattern of features

3. True ☒ False ☒   Tolerances defined in the application of parallelism control, include two parallel lines, two parallel planes, perpendicular planes and cylindrical zones.

4. True ☒ False ☒   The distance between two flat surfaces can be controlled by specifying a parallelism tolerance.

5. True ☒ False ☒   When parallelism is applied to a hole, it is understood that a material condition modifier (specified, or automatic, per Rule #2) must be applied on the tolerance value.

6. True ☒ False ☒   Parallelism may be applied to surface elements only.

Problem #67 — Perpendicularity

Complete before advancing to the solution slide.

1. True ☒ False ☒ When applied to a surface that is shown on the drawing to be normal to the datum feature, a perpendicularity specification controls all surfaces that are likewise shown to be normal to that datum feature.

2. When perpendicularity is applied to the width dimension of a slot or groove, _____ is/are controlled by the specification.

    a. ☒ the centerline, established by the two sides of the slot or groove,

    b. ☒ the surface on the slot or groove that is on the top or the right side

    c. ☒ both sides of the slot or groove

    d. ☒ None of the above.

Problem #67 — Perpendicularity

Complete before advancing to the solution slide.

1. True ☒ False ☒   When applied to a surface that is shown on the drawing to be normal to the datum feature, a perpendicularity specification controls all surfaces that are likewise shown to be normal to that datum feature.

2. When perpendicularity is applied to the width dimension of a slot or groove, _____ is/are controlled by the specification.

    a. ☒ the centerline, established by the two sides of the slot or groove,

    b. ☒ the surface on the slot or groove that is on the top or the right side

    c. ☒ both sides of the slot or groove

    d. ☒ None of the above.

3. True ☒ False ☒   Perpendicularity, as a geometric tolerance, cannot reference more than one datum in an individual feature control frame.

4. True ☒ False ☒   Perpendicularity, when applied to a feature surface, also limits straightness elements, and surface flatness.

5. True ☒ False ☒   When a perpendicularity tolerance is used to show the relationship between two surfaces, a basic angular dimension of 90° should be included.

6. True ☒ False ☒   A surface feature that is controlled by a perpendicularity specification, must be within the perpendicularity tolerance zone *and* the size tolerance.

7. True ☒ False ☒   When applied to a surface feature, perpendicularity is subject to the stipulations of Rule #1—perfect form at MMC.

Problem #68 — Orientation Tolerances

How many of the total number of geometric tolerances are designed to control orientation specifically? _____

Name and sketch the symbol of the geometric tolerances that specifically control orientation.

If two surfaces are dimensioned with a tolerance between them of ± .010, what would be the permissible variation from true parallelism? _____

Surfaces controlled by orientation tolerances must have a *form* tolerance that is ($\leq$ / $\geq$) the orientation tolerance. (Circle the correct response.)

An orientation control that is other than perpendicular or parallel to the controlling datum is controlled with _____.

When an orientation control is other than perpendicular or parallel, the specific dimension for the angle must be _____.

What would be the virtual condition for a hole that has a perpendicularity tolerance of .010, and the hole diameter specification is .562 + .004, - .002? (Show your work.)
_____

Problem #70 — Orientation Tolerances

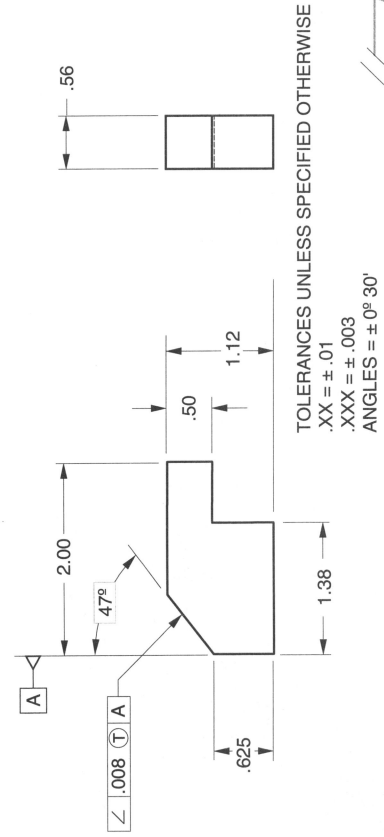

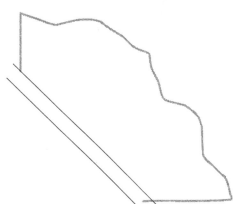

TOLERANCES UNLESS SPECIFIED OTHERWISE
.XX = ± .01
.XXX = ± .003
ANGLES = ± 0º 30'

The encircled "T" in the feature control frame is a _____ modifier.

This type of modifier can only be used with angularity. T ☐   F ☐

The tolerance zone for the geometric angularity control is shown on the drawing to the right. Complete the sketch by adding all necessary lines to show a surface that lies partially outside the specified tolerance zone.

Tolerances of Location
Problems

The following problems correspond with the slides in the *Location Tolerances* presentation. Follow the instructions for each problem. For all of these problems there are solution slides provided in the presentation.

Problem #70 — Orientation Tolerances

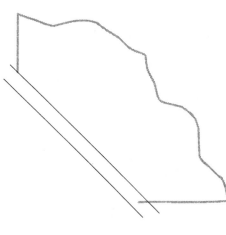

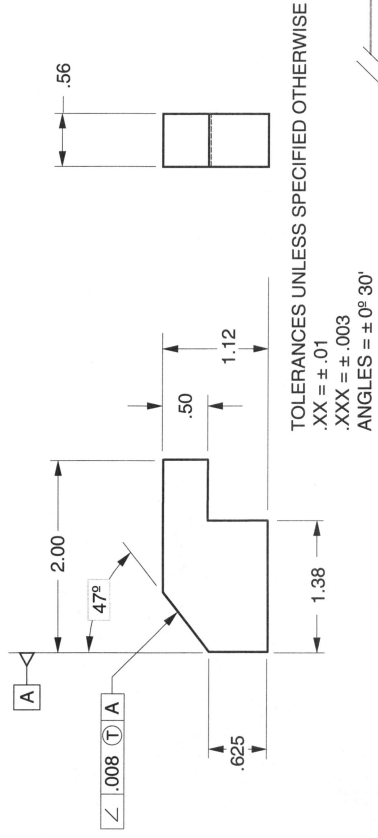

.56

1.12

.50

2.00

47°

1.38

.625

A

∠ | .008 | Ⓣ | A

TOLERANCES UNLESS SPECIFIED OTHERWISE
.XX = ± .01
.XXX = ± .003
ANGLES = ± 0° 30'

The encircled "T" in the feature control frame is a _____ modifier.

This type of modifier can only be used with angularity. T ☐   F ☐

The tolerance zone for the geometric angularity control is shown on the drawing to the right. Complete the sketch by adding all necessary lines to show a surface that lies partially outside the specified tolerance zone.

Tolerances of Location Problems

The following problems correspond with the slides in the *Location Tolerances* presentation. Follow the instructions for each problem. For all of these problems there are solution slides provided in the presentation.

Problem #71—Location Tolerances

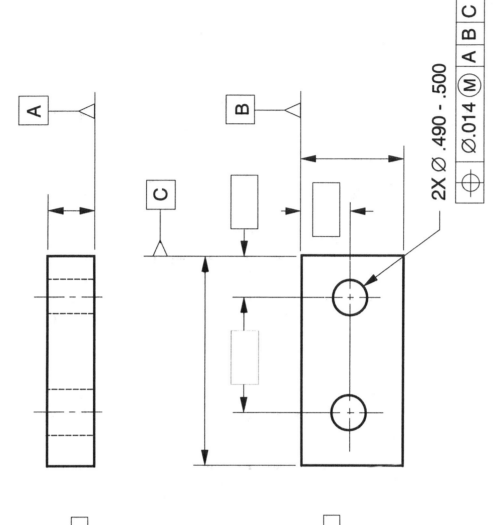

2X ⌀ .490 - .500

| ⌖ | ⌀.014 Ⓜ | A | B | C |

Where position tolerances are used, a diameter symbol must always be included in the tolerance portion of the feature control frame. True ☐ False ☐

In the example drawing, the feature control frame provides for (shift tolerance / bonus tolerance).

The feature control frame is incomplete because there should be a modifier on the primary datum. True ☐ False ☐

What would be the shape of the tolerance zone for the holes?

What controls the orientation of the axes of the holes?

Problem #72—Location Tolerances

Determine the minimum distance for "X". Show your work and explain how datum B would be established and verified.

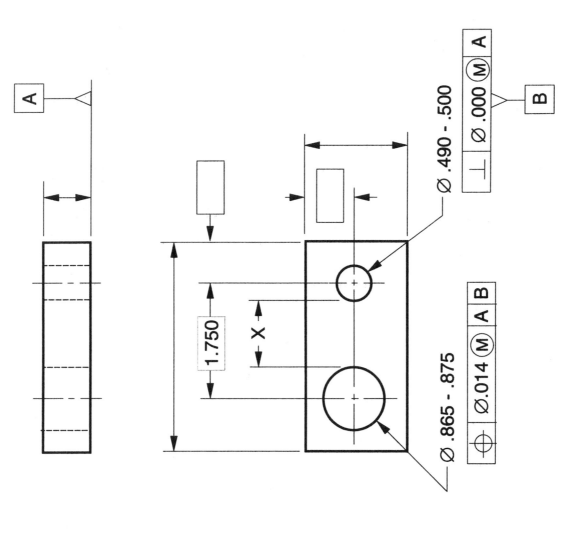

Problem #73—Location Tolerances

Determine the *maximum* distance for "X". Show your work.

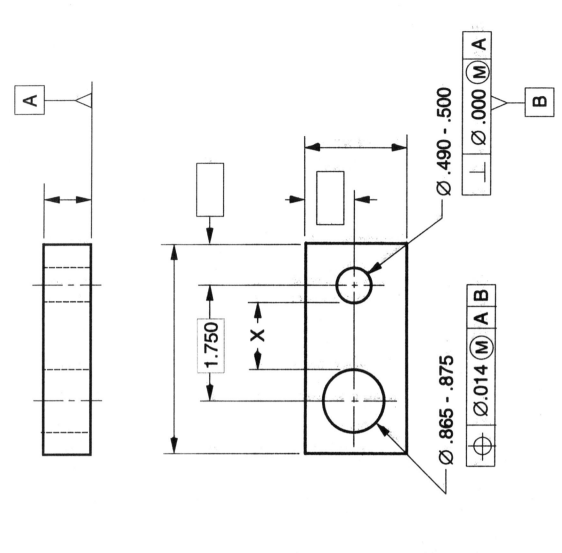

Problem #74—Location Tolerances

Calculate the *maximum* distance
for X. Show your work.

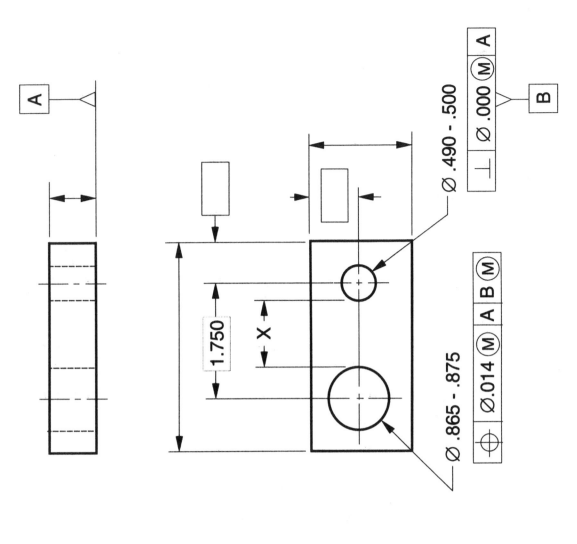

Problem #75—Location Tolerances

Calculate the *minimum* distance
for X. Show your work.

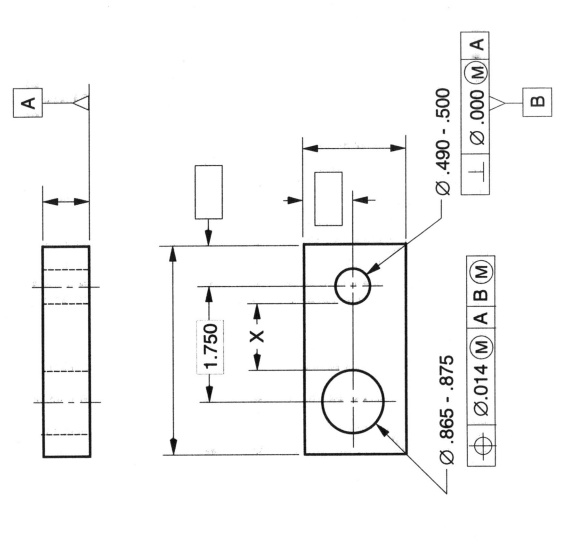

Problem #76—Location Tolerances

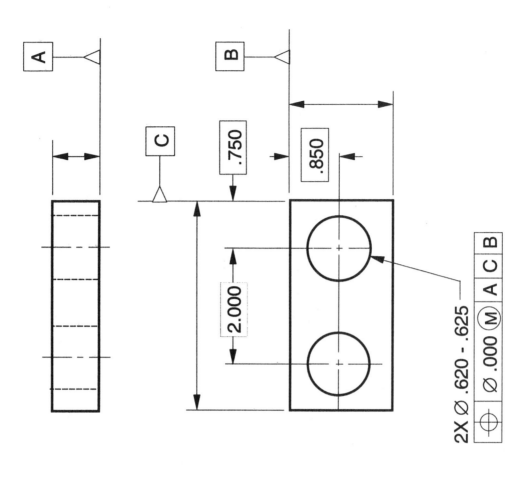

2X ⌀ .620 - .625

⊕	⌀ .000 Ⓜ	A	C	B

Consider the drawing to the right. Is bonus tolerance possible? _____

If so, what is the limit of the bonus tolerance? _____ If bonus tolerance is not applicable, why is that so? _____

What does the basic dimensions actually define? _____

At MMC, what is the location tolerance for the two holes? _____

If the actual measured hole size was .623, what would be the location tolerance? _____

What controls the orientation of the axes of the holes?

Problem #77—Location Tolerances

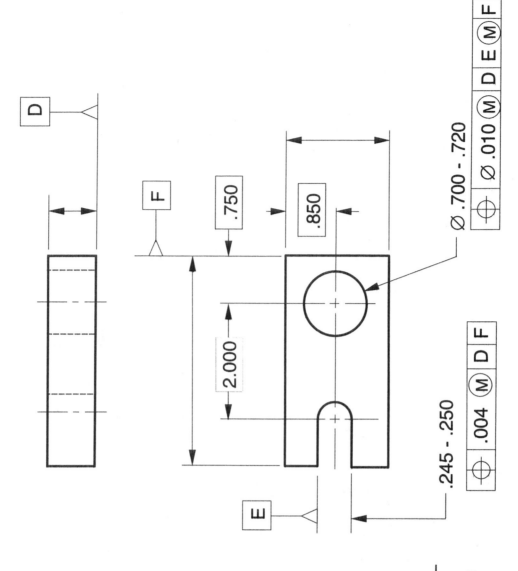

What is the location tolerance on the slot, when the actual size is at LMC? _____

What is the location tolerance for the hole, when the measured diameter size is .713 and datum feature E is at MMC? _____

What is the shape of the tolerance zone for datum feature E? _____

What is the tolerance on the 2.000 dimension between the centers of the two features? _____

Interpret the tolerance zone for the hole.

Tolerances of Profile Problems

The following problems correspond with the slides in the *Profile Tolerances* presentation. Follow the instructions for each problem. For all of these problems there are solution slides provided in the presentation.

Problem #78—Profile of a Line

APPLICATION QUIZ

1. Basic dimensions must be used to locate and define the true profile.

   True ☐   False ☐

2. The geometric profile control must be applied in one of the primary views (Front, Top, or Right Side).

   True ☐   False ☐

3. Modifiers (MMC, LMC) may be used in the feature control frame when controlling profile of a line.

   True ☐   False ☐

4. The profile tolerance must be a refinement of any non-basic locating coordinate tolerance.

   True ☐   False ☐

5. Datums may be used in the feature control frame.

   True ☐   False ☐

6. When profile is used to control form, the tolerance value must be a refinement of the size tolerance.

   True ☐   False ☐

Problem #80—Profile of a Line

Answer the questions below.
Then check your work against the
next slide in the presentation.

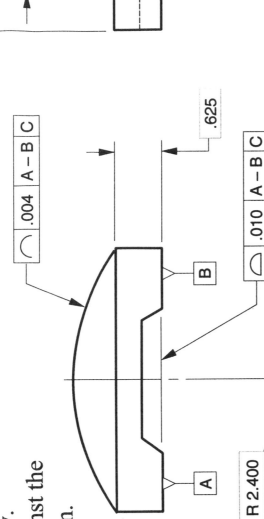

.004 | A – B | C

.625

B

.010 | A – B | C

2 SURFACES

A

R 2.400

C

Which is the primary datum A ☐, B ☐, or C ☐? (Check the appropriate block(s).)

How many datum simulating surfaces would be required to verify this part? _____

Referring to the two flat surfaces on the part, is the profile control unilateral or bilateral?

_____

The two surfaces on the bottom of the part have a flatness limit of _____.

Why is the rounded surface on top located with a basic dimension? _____

_____

Describe the tolerance zone for the profile control on the top surface. _____

_____

Problem #82—Profile of a Surface

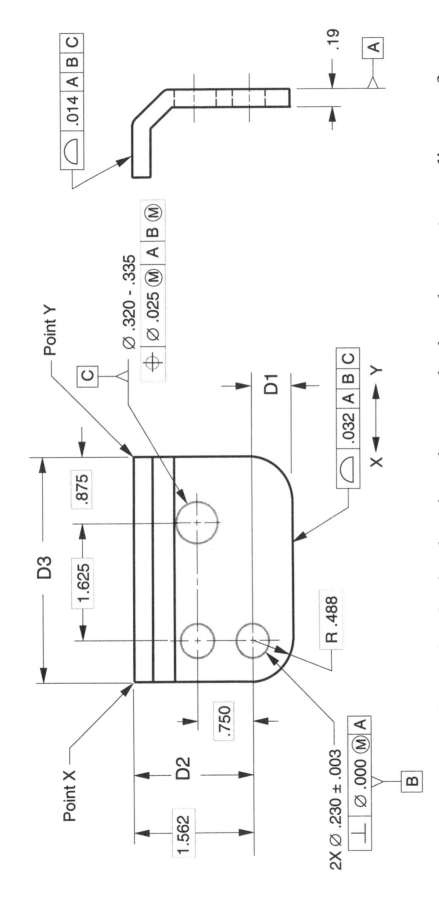

Based upon the information in the drawing above, calculate the *maximum* distance for D1 _____

What would be the minimum distance D1 can be? _____

When the above problems have been solved, advance to the next slide in the presentation.

Problem #84—Profile of a Surface

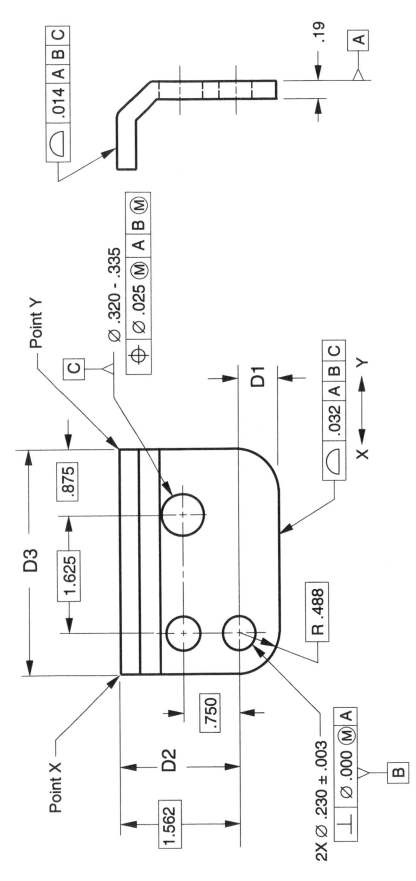

Based upon the information in the drawing above, calculate the *maximum* distance for D3 _____

What would be the minimum distance D3 can be? _____

When complete, advance to the next slide in the presentation.

Problem #85—Circular Runout

Complete this exercise before advancing to the next slide in the presentation.

1. Circular runout tolerances may be applied with or without datum references.   True ☒   False ☒

2. Virtual condition can be affected by circular runout tolerances.   True ☒   False ☒

3. Circular runout tolerances can be applied to features only.   True ☒   False ☒

4. Modifiers may be used with runout tolerances when applied to features.   True ☒   False ☒

5. Datum axes may be established in only one way when using circular runout.   True ☒   False ☒

6. Circular runout permits the application of bonus tolerances.   True ☒   False ☒

7. Runout may be verified with a fixed functional gage.   True ☒   False ☒

8. The concept of shift tolerances is applicable with circular runout tolerances.   True ☒   False ☒

9. When verifying circular runout, the gage must be maintained parallel to the datum axis.   True ☒   False ☒

10. Circular runout is a composite control affecting the form and location of all circular elements simultaneously.   True ☒   False ☒